ROADSIDE GEOLOGY

of Vermont and New Hampshire

Bradford B. Van Diver

MOUNTAIN PRESS PUBLISHING COMPANY
Missoula, 1987

ROADSIDE GEOLOGY SERIES
Editorial Directors:
David Alt and Donald Hyndman

2nd Printing — June 1988

Library of Congress Cataloging-in-Publication Data

Van Diver, Bradford B.
Roadside geology of Vermont and New Hampshire.

Bibliography: p.
Includes index.
1. Geology — Vermont. 2. Geology — New Hampshire.
I Title.
QE141.V36 1987 557.42 87-3897
ISBN 0-87842-203-X (pbk.)

MOUNTAIN PRESS PUBLISHING COMPANY
P.O. Box 2399
Missoula, MT 59806
(406) 728-1900

To Thor, Kent, David, and Mark,
and those they love

Table of Contents

Preface

This book is for everyone who ever traveled in the lovely states of Vermont and New Hampshire and wondered the meaning of their landscapes and tortured rocks. It is a book as free of jargon as I could make it; and for the technical terms that are used, there is a glossary in the back for quick reference, as well as extensive introductory material at the front.

This book is for the traveler, and as such it concentrates on what you can see from a car in motion. I urge you, however, to stop when you can to look more closely at things that pique your interest. The best geology is the rock you can get close to, touch, or even sample. This is outdoor country, filled with natural wonders and lots of trails leading to them. Take a little walk, visit these places, try your hand at being a field geologist, and you will magnify your enjoyment.

A work of this type necessarily depends on the work of many others. All I have done, in fact, is to distill the geological literature, extracting from it what I thought would be most interesting and visible to you, the reader. My wife, Bev, and I traveled all the roads included in the roadguides, and more; and I wrote about what could be seen and what it meant in the context of the body of literature. We used the same system of logging that we did for my earlier book, *Roadside Geology of New York* (1985); I did the driving and observing while she did the recording. As before, Bev made many keen observations too, which were most helpful. Because she has no training in geology, she saw things with a layman's eyes. Her contribution was even greater in the writing stage for, as a teacher of English and a superb writer, she was quick to spot unclear passages, poor sentence structures, and other errors that might trouble the reader. So I made changes. This is her book almost as much as it is mine, and I am most fortunate to have her as companion and co-worker.

Special thanks go, as well, to my colleague, Jim Carl, who critically read the Plate Tectonics chapter and suggested many helpful changes.

The complex geological story written in the rocks of these two states proved difficult to boil down. The landmark paper, "Tectonic Synthesis of the Taconian Orogeny in Western New England," by Rolfe Stanley and Nicholas Ratcliffe, published in the 1985 Geological Society of America bulletin provided the large-scale plate tectonic framework I needed; it is a colossal framework of opening and closing oceans, volcanism, mountain-building, and colliding continents. Adopting this approach allowed me to skip a lot of cumbersome details and concentrate on the highlights and the larger meaning of the bedrock features you, the reader, can actually see in the field. It is to Stanley and Ratcliffe that I owe the greatest debt of gratitude for the bedrock story.

My information about landscapes and their meaning, especially in regard to glacial erosion and deposition during the Ice Age, an important aspect of New England geology, comes from numerous sources. Most notable are the works of David P. Stewart and Paul MacClintock, including their Surficial Geologic Map of Vermont; J. W. Goldthwait; R. P. Goldthwait, L. Goldthwait, R. F. Flint, C. H. Hitchcock, and Ernst Atnevs, to mention a few.

As for other sources, I have relied heavily upon the state geologic maps (Vermont 1961 and New Hampshire 1955) and numerous publications of the Vermont Geological Survey and New Hampshire Department of Resources and Economic Development. Marland P. Billings's monumental 1955 summation, "The Geology of New Hampshire, Part II—Bedrock Geology," was invaluable. I also used papers published by the Geological Society of America and the U.S. Geological Survey, including topographic quadrangle maps of the latter. Last, but not least, I gleaned much useful information from the guidebooks of the New England Intercollegiate Geological Conference. This organization, like its counterpart, the New

York State Geological Association, meets at a different location each fall, mainly for the purpose of conducting local field trips for which the host institution prepares a field trip guidebook. The many guidebooks that have accumulated over the years constitute a most valuable resource. You will not see all of these authors cited in the text in endless procession because that doesn't seem appropriate in a work of this type. Nevertheless, they built the foundation upon which this guidebook stands and I am most grateful for their many contributions.

With only a few exceptions, all of the photographs and line drawings are my own. Photographs of the granite quarry were donated by the Rock of Ages Corporation of Barre, Vermont; and the underground marble quarry by Vermont Marble Company of Proctor. A few drawings came from the literature as noted in their captions. Two marvelous pen-and-ink creations, one of the exposed Champlain thrust near Burlington and the other of Franconia Notch, are the artwork of my former geology student, Terry Jancek.

A colored geologic map accompanies each of the roadguides, showing the route and some of the villages in black. In most cases, this map is all you will need to find your way around, but since only the principal road, or roads, are shown on it, it would be helpful to have a state road map, too. I described the routes going from A to B, but tried to do so in a way that makes it easy for anyone to go from B to A.

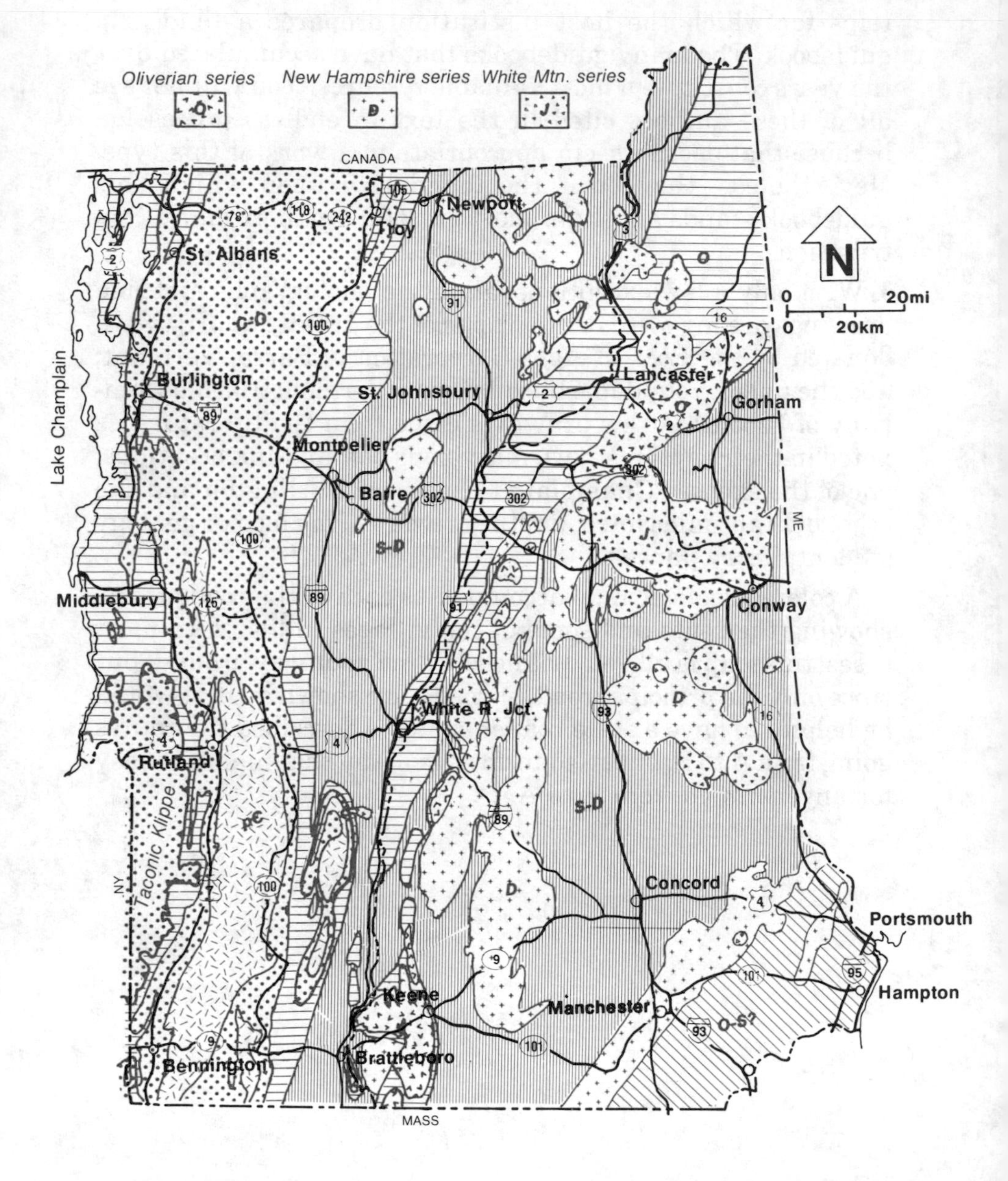

Simplified bedrock geologic map, omitting faults, with routes covered in the roadguides.

Minerals, Rocks, and Geologic Time

Every profession has its own distinctive language. As in other highly specialized fields, it is nearly impossible to talk about geology without first defining some terms and giving a little background. Accordingly, this section is designed for the layman with no geological training.

MINERALS

Minerals are the basic building blocks of the solid earth. Minerals are all around you—in the rock, in the soil, in your home. Sheetrock in the walls of your home, for example, is made of the mineral gypsum. The salt you sprinkle on your eggs is the mineral halite, rock salt. Snow is another mineral, ice. Residents of Vermont and New Hamphshire know it well.

What is a mineral?

Geologists apply the term only to natural solids with limited variability of composition and internal structure that possess characteristic physical properties such as color, hardness, and shape.

Crystal Structure

Table salt, or halite, composed of equal parts of the elements sodium and chlorine, can be identified by its salty taste. But look at some table salt under a magnifying glass and see its

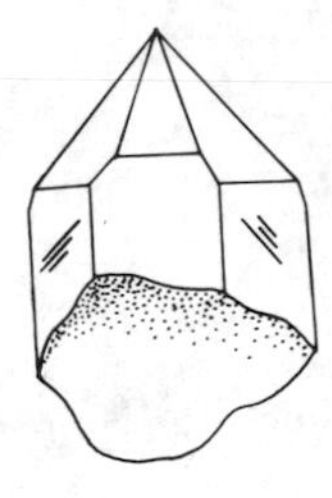

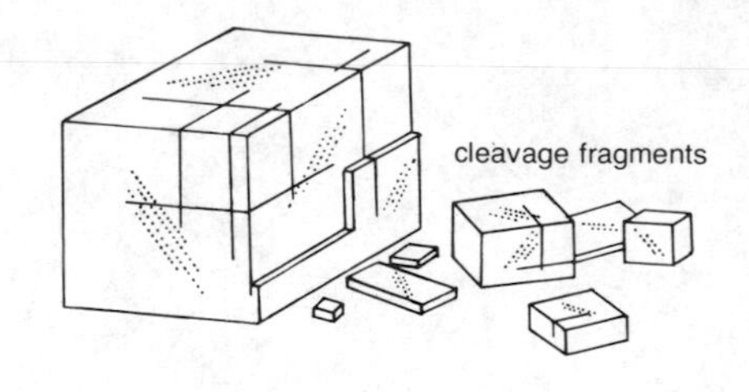

On the left, a broken quartz crystal illustrates crystal structures without cleavage. On the right, the mineral halite, or common table salt, splits easily along 3 cleavages at right angles to each other, an external expression of its internal cubic framework of sodium and chlorine atoms, its crystal structure.

cubic crystals. Every time halite is crushed, it breaks up into cubic fragments, even when ground to a powder. The halite cubes tell us that the sodium and chlorine atoms are arranged in a cubic framework and that there are three planes or cleavages at right angles to each other along which the framework separates freely. Other minerals have different frameworks and different cleavages. Some, like quartz, may form crystals with smooth faces but have no cleavage, and tend to break into irregular fragments.

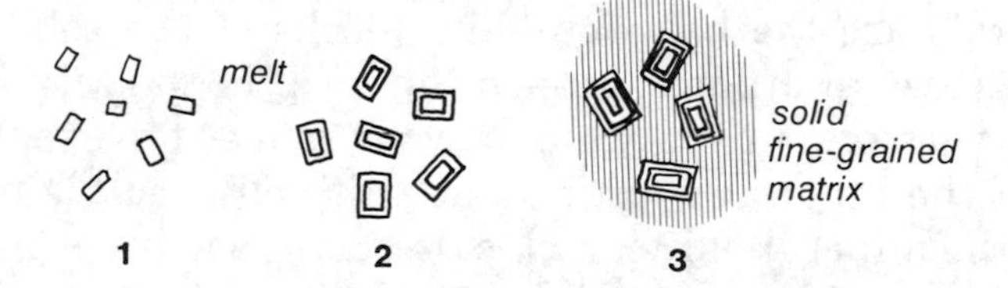

"Floating" crystals in porphyritic rocks as in (3) above, show how minerals form by igneous processes; in growth stages (1) and (2), they grow freely in magma; in (3) the remaining magma has solidified rapidly, as when it is erupted and quickly cools.

How do minerals form?

Lava is made of atoms and molecules free to move more randomly. If cooled the atoms and molecules combine into crystalline frameworks; this is one of the ways minerals form—by slow cooling of molten rock.

A second mode of origin is from water solutions, which takes

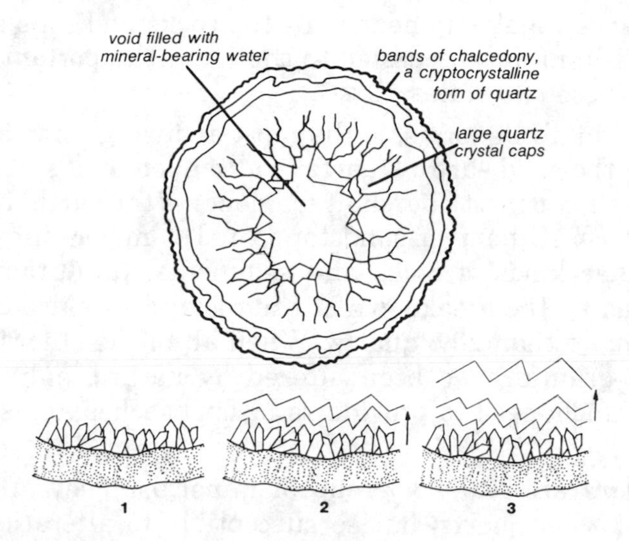

Growth of crystals in a geode begins with random small crystals (1). As the crystals enlarge and interfere, some die out as in (2) and (3), so that the final crust includes a relatively small number of shiny caps.

us back to the table salt. The salt you buy comes from deposits precipitated from very salty water, such as Utah's Great Salt Lake. All seawater contains dissolved sodium chloride, but its high solubility and lower concentration keep it in solution. In some places, however, like Great Salt Lake or the lagoons along the Texas Gulf Coast, high evaporation rates and limited inflow concentrate the salt and cause it to precipitate from solution as the mineral halite. Other examples of minerals precipitating from solution are the calcite of stalactites found in some caves, and the cement deposited in the pore spaces of sand to form sandstone.

The third important mode of mineral origin is metamorphism, the recrystallization of existing minerals under high temperatures without melting. It involves the migration and reorganization of atoms and molecules in solid rock. Metamorphic and igneous minerals are most important in Vermont and New Hampshire because almost all of the bedrock is metamorphic or igneous.

Common Rock-forming Minerals

Geologists recognize hundreds of minerals, most of them

rare. Just ten make up nearly all the rocks and soils around you. Let's limit our discussion to those most important in the rocks of these two states.

Quartz, in all its forms, is the most enduring common substance of the solid earth. Quartz is a compound of silicon and oxygen, the two most abundant elements of the earth. It is the principal constituent of sandstone, and a major mineral in many other kinds of rocks and sediments. Its durability is illustrated by the breakdown of the rock granite, which may contain more than 20% quartz. When all of the other components of granite have been altered by nature, only quartz remains unchanged to provide sand for beaches and streambeds.

Feldspars are really a group of minerals. They are more abundant than quartz; but so susceptible to alteration that they are soon converted to clays. Feldspars contain aluminum, silicon, and oxygen. Two kinds, orthoclase and microcline, often called potash feldspars, also contain potassium and some sodium. Plagioclase feldspars contain sodium and calcium in variable proportions. Feldspars are major constituents in a wide range of igneous and metamorphic rocks, but relatively minor in most sediments because of their lack of durability.

Micas occur principally as dark *biotite* and clear *muscovite*. Both are perfectly cleavable into very thin, glossy sheets. This property and resistance to heat make some mica useful for the small windows used in some laboratory furnaces. Mica is principally silicon, oxygen, aluminum, water, and potassium. Biotite also contains variable amounts of iron and magnesium which make it dark.

Amphiboles are a group of closely related minerals. All contain silicon, oxygen, and water, but beyond that, their composition varies widely. The most common member, hornblende, also contains aluminum, iron, magnesium, calcium, and sometimes titanium. Geologists often refer to it, as a "garbage pail mineral." A very common constituent of both the igneous and metamorphic rocks of Vermont and New Hampshire, hornblende can be easily recognized by its dark green to black color and two perfect cleavages at angles of 126 degrees and 54 degrees, giving a diamond-shaped cross section. The cleavage angles are the same for all amphiboles.

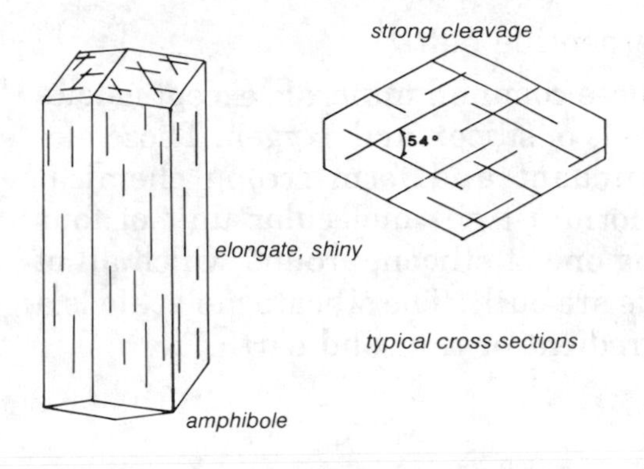

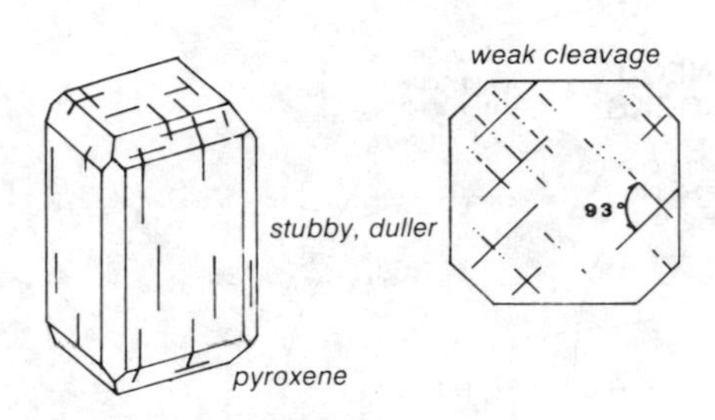

Hornblende and pyroxene can often be distinguished by some of the properties shown

Pyroxenes are a group of minerals that resemble the amphiboles in many ways. They abound in the serpentine belt of Vermont. Like hornblende, most pyroxenes are dark colored, but can be distinguished from the amphiboles by two imperfect cleavages which intersect each other at close to 90 degrees.

Calcite and *dolomite* are the principal minerals of the rocks limestone, dolostone and marble. They are difficult to tell apart. Both can be scratched by a pocket knife blade and both have perfect cleavage in three directions not at right angles to each other—so as to form a rhombohedron, like a pushed-over cube.

Clay minerals are a major constituent of many types of sediments, soils, and rocks. They are compounds of silicon and oxygen with a wide range of chemical composition. All are so extremely fine-grained, that geologists must use the electron microscope to see individual grains. The clay-rock called shale can usually be recognized by both its fine grain size and its thin laminations produced when clay mud is compressed into rock.

Olivine contains silicon and oxygen along with iron and magnesium in varying proportion. Like pyroxene it is an important mineral of the serpentine belt.

As you can see, all of these common minerals except calcite and dolomite are compounds of silicon and oxygen. Those elements are extremely abundant and form strong chemical bonds. They combine to form a tight molecular unit of four oxygen atoms surrounding one of silicon, around which all of the many silicate minerals are built. The silicate molecule is a very important basic ingredient of the solid earth.

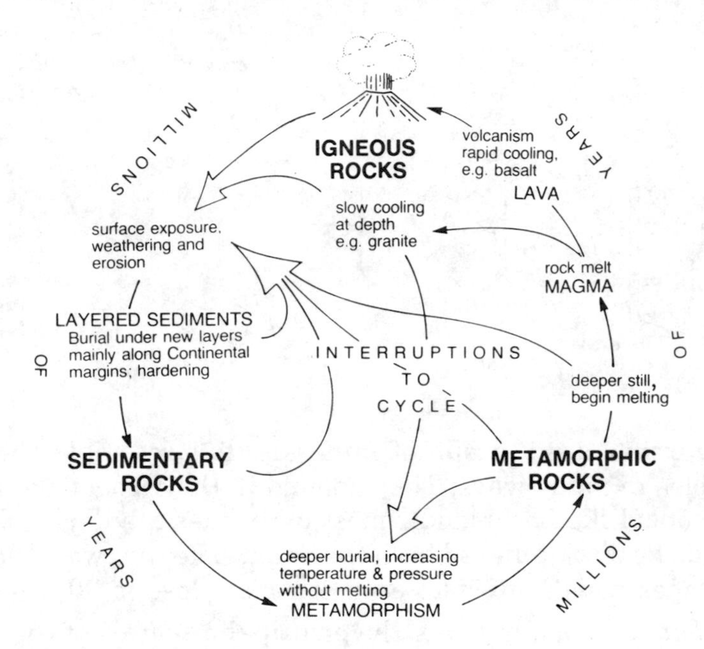

The rock cycle, showing the three major classes of rocks, how they form, and how they interrelate

ROCKS

Rocks are aggregates of minerals. Most contain several minerals, but some, like rock salt or pure limestone, are composed of just one kind. Geologists distinguish three fundamental groups of rocks: igneous, sedimentary, and metamorphic. The rock cycle illustrates how they form and interrelate.

Igneous Rocks

Igneous rocks form as molten rock, magma, crystallizes. Crystallization of different magmas produces a wide range of intrusive, or plutonic, igneous rocks such as granite. Intrusive rocks tend to be coarse-grained because buried magma cools slowly enough for large crystals to grow. If magma reaches the surface, it becomes lava. Crystallization of different lavas produces a wide range of volcanic igneous rocks such as basalt. Volcanic, or extrusive rocks tend to be fine-grained because they crystalize rapidly.

Granite is a coarse-grained, light-colored rock, composed principally of feldspar and quartz. Some, like the Conway granite, are pink because their potassium feldspar is pink. The mineral grains are homogeneously distributed and lack alignment, except near contacts. The intrusive liquid origin of granite is obvious where there are inclusions, called xenoliths, chunks of country rock that fell into the magma and were imprisoned when it solidified. These are most abundant near contacts.

Syenite resembles granite, but is dominantly potassium feldspar. It contains little or no quartz or plagioclase feldspar.

Quartz monzonite, granodiorite, quartz diorite differ from granite principally in their different proportions of potash feldspar and plagioclase feldspar, which are hard to judge without a microscope. Referring to them all as granite is not too far off the mark. All tend to be darker than granite. The Kinsman quartz monzonite is one of the most striking rocks of New Hampshire. It is mostly a rather dark, biotite-rich rock with xenoliths of all sizes of darker country rock. It commonly contains rectangular, white potassium feldspar crystals up to about 3 inches long.

Diorite and *gabbro* are dark intrusive rocks with virtually no potassium feldspar or quartz. They consist principally of the dark, iron-magnesium minerals pyroxene and hornblende, along with plagioclase feldspar. Both are uncommon.

Ultramafic rocks are very dark-colored rocks largely composed of iron-magnesium minerals such as pyroxene, hornblende, and olivine, or their alteration products, particularly serpentine and talc. The serpentine belt of Vermont contains hundreds of small slivers of these rocks in highly sheared and thrust-faulted schists.

Dikes are thin and steeply inclined igneous bodies formed by intrusion of magma into fractures. They abound in Vermont and New Hampshire, especially around the White Mountain magma series. Basaltic dikes, with their black to dark gray, fine-grained rock and rusty-weathered surface, are the most conspicuous. There are many lighter-colored varieties. One, called aplite, is really a fine-grained, sugary-textured granite.

Pegmatites are very coarse-grained dike rocks, closely related to aplite and also of late magmatic origin. Most are granitic and white. Feldspar crystals are typically several inches long, in some cases much longer. Large books of mica are very common, and the spaces between these minerals are generally filled with glassy quartz. Pegmatites are often sought by mineral collectors because they may contain nice crystals of exotic minerals such as beryl and tourmaline, among many others. Most of the pegmatites of this region are related to intrusions of the New Hampshire plutonic series.

Sedimentary Rocks

Sedimentary rocks are formed by the consolidation or lithification of sediments, most of which form through destruction of older rocks. All rocks exposed above sea level are subject to weathering, so sediments are always forming.

Most sedimentary rocks form on the edges of the world's continents. Practically all of the limestones and dolostones are shallow water deposits of the continental shelf. They are the most distinctive ingredients of the "shelf sequence," which also contains sandstones and shales. Carbonate rocks are rare in deeper water because they dissolve in the colder water. Rocks formed there are mainly shale, graywacke, a kind of dirty sandstone deposited from submarine landslides called turbidity currents, and volcanics. Fossils are the remains or traces of animals and plants preserved in the rocks by natural processes. They are common in limestones and shales, uncommon in coarse-grained rocks like sandstone.

Metamorphic Rocks

Metamorphic rocks form from older igneous, sedimentary, or metamorphic rocks. The transformation, called metamorphism, may be compared with changes in the interior of a glacier. As a glacier moves, the millions of interlocking ice crystals that make up its mass are repeatedly sheared, broken, distorted, and recrystallized without significant melting. The grains are reshaped, compacted, and stretched out. If you were to descend into a glacial crevasse, you would see that the ice is very hard and brittle. Yet, the net result of all these slow granular and intergranular changes is plastic movement of the whole mass, as though it were taffy.

Similar changes occur in hard rock, but under much greater pressure and at higher temperature. Several years ago, I went down 1600 feet into a salt mine near Houston, Texas. The temperature at that depth was 98 degrees Fahrenheit while, on the surface, it was only 70 degrees Fahrenheit! As you go deeper into the Earth, the temperature and pressure both rise.

Most of the metamorphism in the two states is regional metamorphism that affects a large area. Regional metamorphism, like most of the other action in the rock cycle, takes place at the edge of the continent in deeply buried rocks. In general the deeper the burial the more intense the metamorphism. Regional metamorphism typically involves pervasive microscopic shearing and granulation of the rock, while new minerals crystallize parallel to the shear planes. The resulting parallelism of mineral grains is the principal cause of the strong foliation that is the trademark of slates, phyllites, schists, amphibolites, and gneisses, among other metamorphic rock types.

Metamorphic temperatures may also exist near igneous intrusions. Their effect upon the intruded rocks, called contact metamorphism, generally does not extend far from the intrusion. Unlike regionally metamorphosed rocks, these rarely show much distortion or foliation except what was already there. Composition may, however, be drastically changed as mineral matter moves from the magma into the intruded rock.

Certain minerals found in metamorphic rocks give clues to the metamorphic grade, which depends mostly on temperature, but also on pressure. These "index minerals" form in host rocks of appropriate composition under certain conditions of temperature and pressure. Because the index minerals for rocks formed from shales are most important and referred to repeatedly in the following road guides, they are listed on the next page for easy reference.

Here are some of the more common metamorphic rocks of Vermont and New Hampshire.

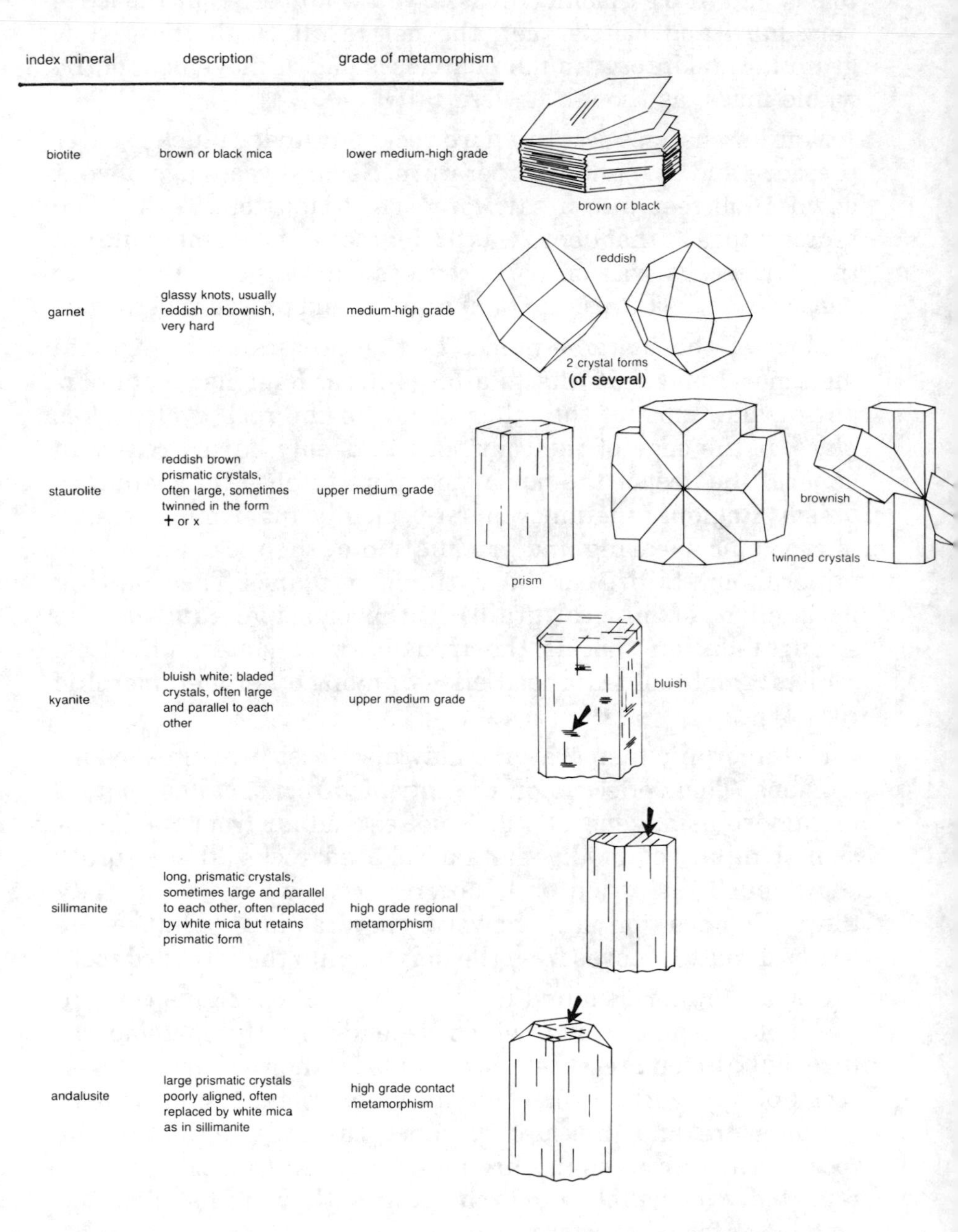

index mineral	description	grade of metamorphism
biotite	brown or black mica	lower medium-high grade
garnet	glassy knots, usually reddish or brownish, very hard	medium-high grade
staurolite	reddish brown prismatic crystals, often large, sometimes twinned in the form + or x	upper medium grade
kyanite	bluish white; bladed crystals, often large and parallel to each other	upper medium grade
sillimanite	long, prismatic crystals, sometimes large and parallel to each other, often replaced by white mica but retains prismatic form	high grade regional metamorphism
andalusite	large prismatic crystals poorly aligned, often replaced by white mica as in sillimanite	high grade contact metamorphism

Some of the metamorphic index minerals may often be identified without a microscope.

Marble is a generally pale rock formed by metamorphism of limestone or dolostone. In addition to its light color, it can often be recognized in roadcuts by its cavities, formed where openings have been widened by solution in slightly acid water. The cavities commonly appear darker than the fresh rock. Marble is soft enough to be easily scratched with a pocket knife. Because it weathers easily, marble is most often found in valleys.

Quartzite is metamorphosed sandstone that is generally harder, glassier, and feels less sandy. It is a tough, erosion-resistant rock that forms ridges. Original bedding is commonly well-defined and color tends to reddish or brownish shades or white. Quartzite is a brittle rock that shatters into jagged fragments when blasted for roadcuts.

Slate is a fine-grained, low-grade metamorphic equivalent of shale that splits easily into thin, flat plates along cleavage planes. It is found in abundance in many parts of both states and is quarried from the Taconic Mountains of southwestern Vermont and adjacent New York for use as roofing and floor tile. It comes in a wide variety of generally dark colors and forms crumpled, slabby outcrops and roadcuts.

Phyllite is a slightly higher grade equivalent of slate, most easily distinguished by a pearly luster on cleavage surfaces, which also tend to be crinkled and uneven. The luster comes from a surface mat of very fine-grained white mica crystallized during the metamorphism.

Schist is a general term for rocks of a wide range of mineralogical makeup and the name is usually prefixed with those mineral names. Their common feature is strong layering of platy or elongate metamorphic minerals that often permits them to cleave easily.

Greenschist is a low-grade schist containing abundant green minerals, including chlorite soft enough to be scratched by a fingernail. Most of the many greenish roadcuts on Interstate 89 between Burlington and Montpelier are not true greenschists but chlorite-bearing biotite schists. Greenschists typically form by regional metamorphism of basalts.

Mica schist is the most common type of schist. It is the coarser-grained equivalent of slate and phyllite formed by medium to high-grade regional metamorphism of shales. In outcrop, the rock often appears brownish because of the abun-

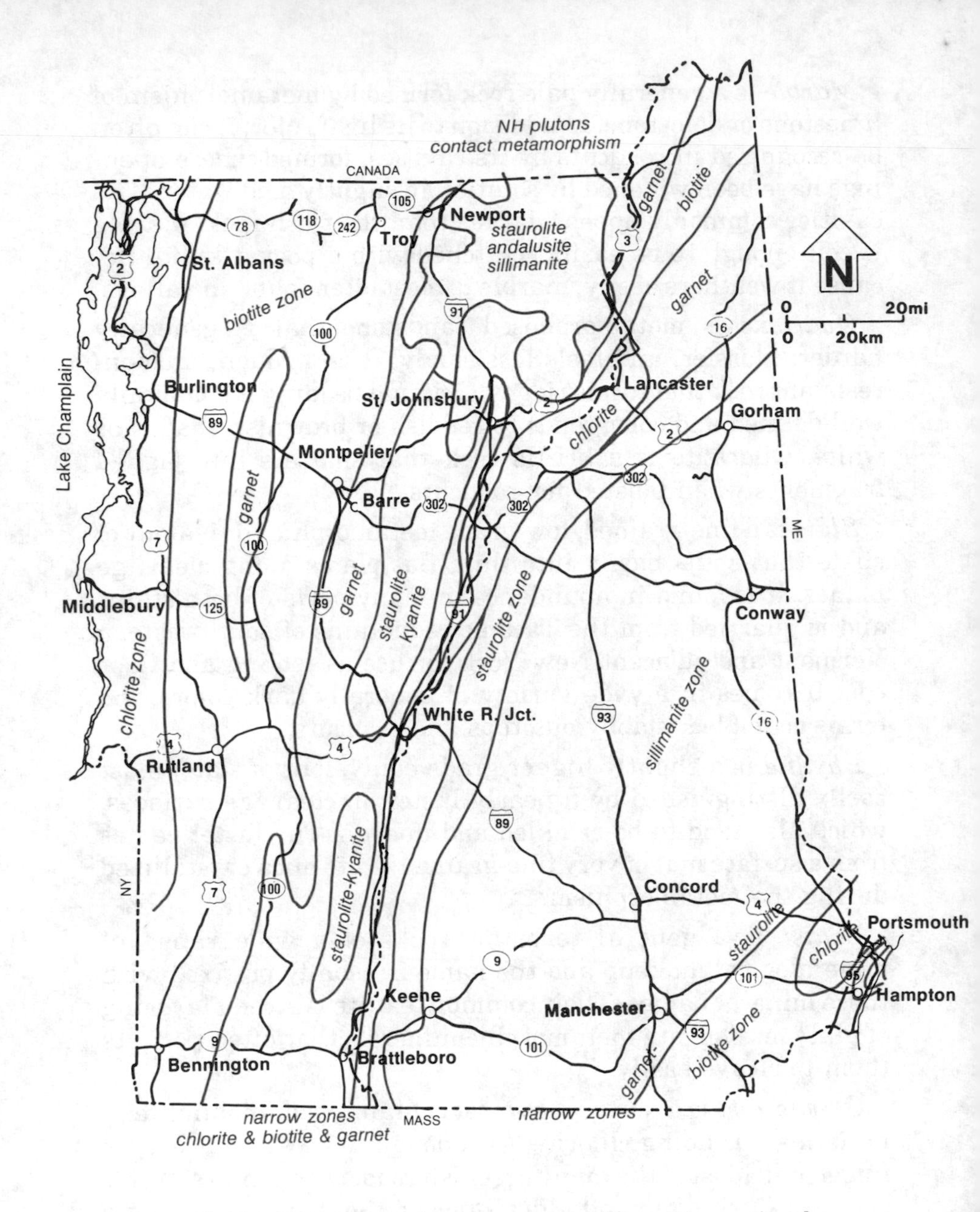

Generalized metamorphic mineral zone map, illustrating regional, mountain-building metamorphism, locally imprinted with higher-grade contact metamorphism resulting from contemporaneous intrusion of New Hampshire series plutons during Acadian mountain-building. Most of the later White Mountain series plutons intrude high-grade rocks which are minimally affected by the intrusive heat. Note the general west-east increase in metamorphic grade.

dance of dark biotite mica, and mica flakes may sparkle in the sunlight. Mica schist may also contain the index minerals garnet, staurolite, kyanite, sillimanite, or andalusite. When it does, those names are attached as prefixes.

Gneiss is a coarse-grained, streaky, commonly dark- and light-banded rock formed by regional metamorphism. Like schist, the term is generally applied to rocks of a wide range of mineralogical makeup, and it is usually prefixed with compositional indicators. For example, granitic gneiss contains the same minerals as granite, and hornblende gneiss is one that contains the mineral hornblende. Most gneisses of this region inherit their coarse texture from original intrusive igneous rocks that approximate granite in composition. Others, however, form through extreme metamorphism of shales or other rock types; the texture tends to coarsen with ascending metamorphism, provided the internal shearing is not too severe. Most of the gneisses of Vermont and New Hampshire are light-colored, owing to the abundance of feldspar and quartz.

Amphibolite is a dark green to black schist made up of densely packed hornblende and plagioclase feldspar. It generally forms through medium-grade regional metamorphism of basalt. It commonly contains knots of red garnet.

The bedrock foundation of Vermont and New Hampshire is entirely metamorphic and igneous. Sedimentary rock names such as shale, limestone, or dolostone are applied to rocks exposed in the Vermont segment of the Champlain basin and the Vermont Valley, but even they are slightly metamorphosed. Metasedimentary rocks younger than Precambrian have all been assigned ages based primarily on the fossils they contain. This is rather remarkable because fossils are easily destroyed by metamorphism, especially when it is of the intense, high temperature variety evident in many parts of this region. Wherever possible, fossil evidence found in correlative formations outside the state was also used in the age determination. Radiometric methods were used to date igneous activity and metamorphism.

As you can see, metamorphosed sedimentary rocks really have two ages: the earlier, based on fossils, gives the time of formation of the original rock; and the later, based on the radiometric clock, gives the time of mountain-building, of which the metamorphism is part.

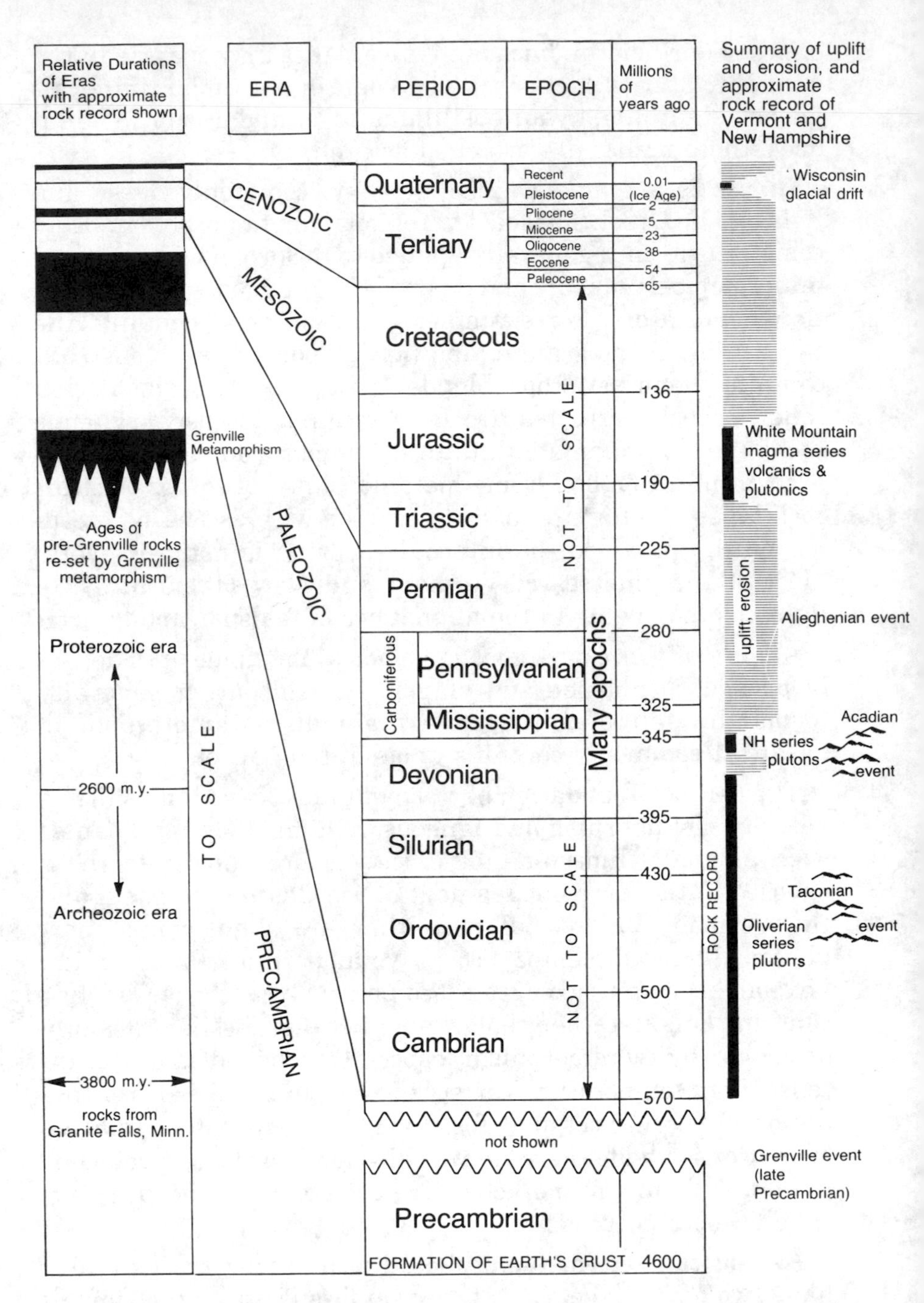

The column on the left is scaled to the full length of Earth's history and the relative durations of the geologic eras, with black bars indicating the preserved rock record. The right column is not to scale, and shows time divisions and tectonic history of post-Precambrian time only.

GEOLOGIC TIME

The ancient geologic history of Vermont and New Hampshire is written in their rocks as surely as if it were printed; and you will be able to read much of the story with a little background. Just being able to identify the rock type and knowing how rocks form reveals half of the story. Rocks are always forming somewhere on earth, albeit slowly, while others are being eroded elsewhere, or melted, or metamorphosed—recycled.

The Geologic Time Scale is the product of this kind of piecing together of rock records, not just from these states, but from the entire world. Obviously, we can only claim a very small part of the total history of earth; information to fill in the enormous gaps, the missing pages, called unconformities, must be obtained from somewhere else where that part of the record is intact. The dating of different rocks in the record has been laboriously achieved by using such evidence as the fossil record, stratigraphic position, and the decay of radioactive elements such as uranium that are sometimes contained in rock-forming minerals. Once dated, the rock units are placed in their proper chronologic order—like the pages of a book. Graphically, the oldest rocks, and the geologic events they depict, are usually placed at the bottom of the scale because that is most commonly where nature has placed them.

The earth is at least 4600 million years old. It is difficult to comprehend such a large span of time; and we can't relate to it with our human clocks, consisting of days, weeks, months, years—even millenia are too short. The geologic clock ticks away a million years at a time, and the Geologic Time Scale fills the need for a different kind of calendar. Its basic units are called periods. Just as our weeks are lumped together in months, the geologic periods are lumped into large units called eras, and just as we divide weeks into days, periods are split into epochs. The divisions are based on significant events recorded in the rocks; such as mountain-building, submersion beneath the sea, or the rise or fall of certain life forms. As a result, none of the divisions represent the same number of years.

As you travel the highways and byways of Vermont and New Hampshire, let the Geologic Time Scale be your road map to history. Throughout the book, you will repeatedly come across names like Precambrian, actually two eras, the older Archean and the younger Proterozoic, Cambrian (period), or Devonian (period). I have avoided the use of epoch names altogether except to refer to the Pleistocene epoch as the Ice Age. I use that term loosely here to mean the whole Pleistocene rather than individual continental glaciations within it. Geologists know these names like the backs of their hands because they must use them constantly as a time frame. If you keep referring to the time scale when you encounter them, you, too, will soon become familiar with them and the events they depict.

Folded block of Devonian Waits River schist in Knox Mountain granite of the Devonian New Hampshire plutonic series, partly "granitized" by injection of magma between the layers and by reaction with the magma. Near Marshfield, Vermont.

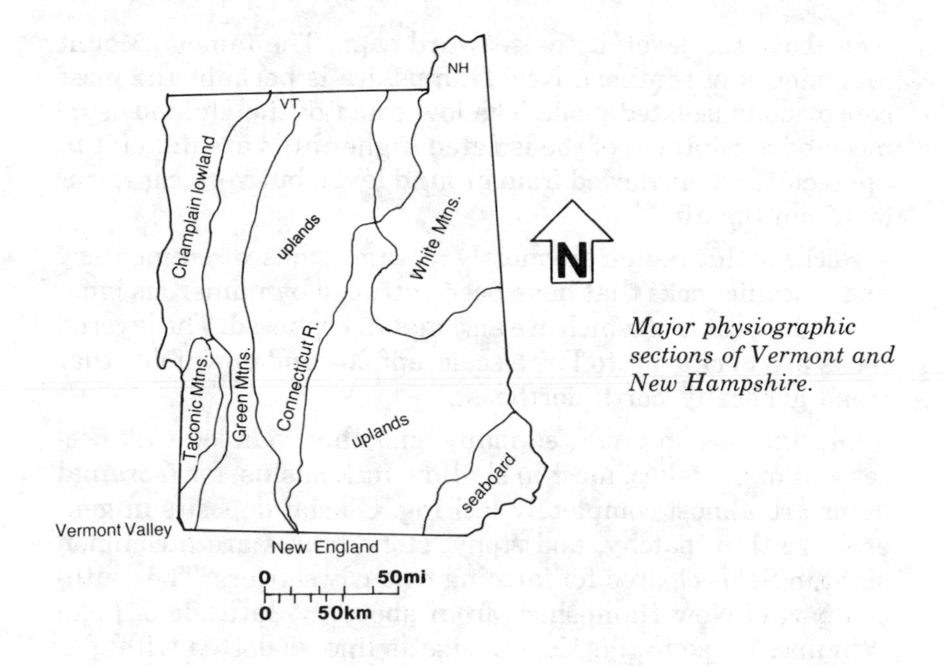

Major physiographic sections of Vermont and New Hampshire.

Looking at the Landscape

Geology and topography are inextricably tied to each other. Knowing something about rocks, landscapes, and the geologic processes that shape the land enables the traveler to see beyond the mere surface features, to know something of what lies beneath, to read geologic history, and to experience the excitement of discovery.

It has long been the practice of geographers and geologists to divide the land into physiographic provinces or sections, each of which is distinguished by certain topographic and, incidentally, geologic characteristics. Overall, Vermont and New Hampshire present a plateau-like landscape called the New England Upland that rises gradually inland from the sea and is capped in places by mountain ranges and individual peaks. Others refer to the Uplands as the Piedmont. The principal ranges are the Taconic and Green mountains in Vermont and the White Mountains in New Hampshire. The Uplands surface slopes southeastward from a maximum of 2200 feet to 400-500 feet above sea level at the seaward edge. The famous Mount Monadnock of southern New Hampshire is perhaps the most conspicuous isolated peak. The low relief of the Uplands surface and abruptness of the isolated higher hills are difficult to appreciate when viewed from ground level, but very conspicuous from the air.

Rocks of this region are mostly metamorphosed sedimentary and volcanic rocks that have been intruded by numerous igneous bodies, some of which are also metamorphosed. The layered rocks are buckled into large-scale upfolds and downfolds that trend generally north-northeast.

Glacial erosion produced many smoothly rounded rock bosses and many lakes, most in shallow rock basins, for morainal dams are almost completely lacking. Glacial deposits in general are thin, patchy, and stony; stone fences are a common sight in fields cleared for farming by early settlers. The southern part of New Hampshire, from about the latitude of Lake Winnipesaukee to the Massachusetts line, is dotted with glacial drumlins that give parts of this section of the Uplands a rolling topography.

Larger valleys of the region, particularly those of the Connecticut and Merrimack rivers, are filled to appreciable depths with glacial drift, principally with outwash and various glacial lake deposits. Much of that stuff has been removed by postglacial downcutting of the rivers, which have left striking stream-cut terraces on the valley sides.

The seaboard section of New Hampshire is a coastal strip of, and slightly lower than, the projected surface of the New England Uplands. It is land that was submerged and eroded by wave action during the waning stages of the Ice Age before the ice-depressed land rebounded but sea level rose nearly to preglacial levels. The lesser altitude is not the result of softer rocks; some, such as the Rye formation, are really very hard.

The White Mountains section includes the Presidential Range, the Franconia Mountains west of them, and the hilly country of northeastern Vermont and northern New Hampshire. Igneous intrusive bodies dominate the region. Many of the mountains are held up by the New Hampshire and White Mountain series granites, but the highest peaks of the Presidential Range are carved from the Littleton schist. All of the main summits in the Presidential Range are well above timberline, about 4500 feet above sea level. The range is capped by 6288-foot Mount Washington, the highest peak in the northeast. Many lesser mountains nearby also have rather barren rocky summits, a characteristic of the granite bedrock.

Mount Washington and other high peaks of the Presidential Range rise from rather flat, high altitude surfaces collectively referred to as the Presidential Upland. Some geologists believe these are remnants of a once-continuous pre-glacial erosional surface. The north, east, and south flanks of the range are deeply scalloped by cirques, locally called "gulfs," or "ravines," carved by small mountain glaciers in the early, and perhaps also late, stages of the last great ice age. Summits throughout the White Mountains section tend to be rounded, with steeper south-facing slopes, a result of erosion by overriding ice. Some north-to-northwest-trending valleys are remarkably ice-gouged; prime examples are Crawford and Zealand notches, southwest of Mount Washington.

The Green Mountains section dominates the Vermont landscape. The range stretches the length of the state, continues into Massachusetts as the Hoosac Mountains and into Quebec as the Notre Dame Mountains. It is about 35 miles wide at the southern end of Vermont, and narrows northward to 20 miles at the Canadian border. Average elevation is about 2000 feet, but many peaks rise above 3000 feet, six exceed 4000 feet, and Mount Mansfield is the highest at 4393 feet. The dominant structure of the Green Mountains, their reason for existence, is an enormous, north-south-trending upfold called the Green Mountain anticline with a backbone of hard Precambrian rocks that is exposed in the southern half of the range. The Precambrian core is everywhere mantled with metamorphosed sedimentary formations of Cambrian and younger age, strata that dip, or slope, eastward and westward away from the crest of the range.

Glacial drift is generally thin in the Green Mountains, except in some of the valleys. The Missisquoi, Lamoille, and Winooski valleys, in particular, contain extraordinary terraces that indicate great depths of original valley fill.

The Taconic Mountains of southwestern Vermont lie west of the Vermont Valley, which separates them from the Green

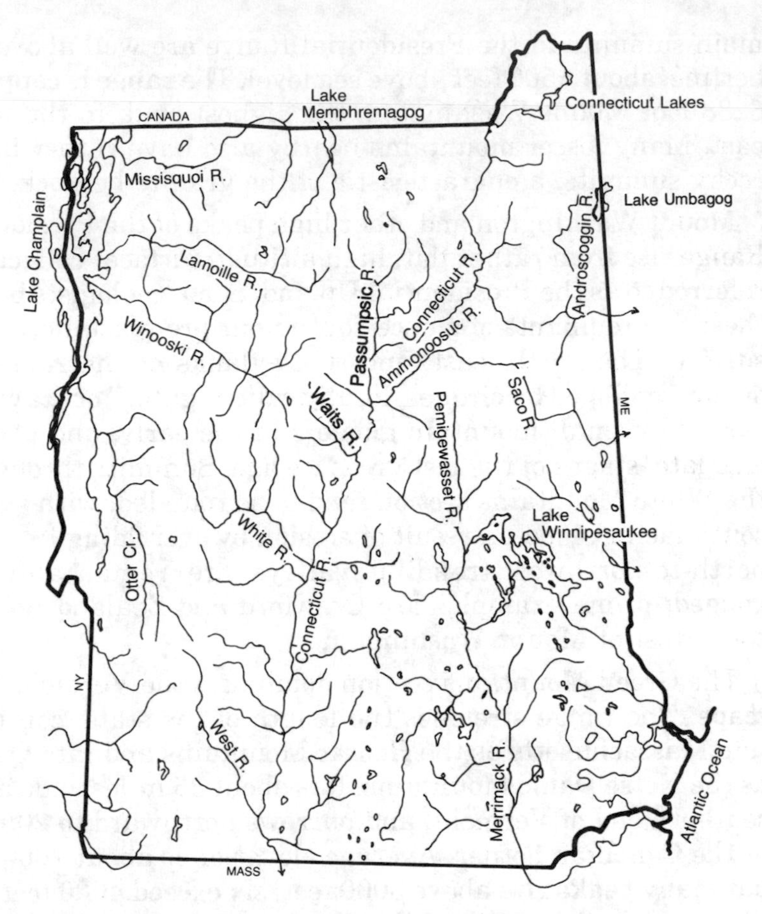

Drainage nets of Vermont and New Hampshire.

Mountains. They are bound on the west by the Hudson-Champlain lowlands, on the north by the Champlain lowlands. The range extends south-southeastward to the Hudson Highlands of New York. Summit elevations are commonly 1800 to 2000 feet, but a few rise to 2500 feet.

Geologically, the range is the Taconic klippe, a large segment of a thrust slice that has been isolated by erosion. Numerous thrust faults chop the range into lesser slices, and the rocks, mainly slabby slates, phyllites, and schists, are extensively folded, cleaved, and sheared. This is reflected in a rather lumpy, irregular landscape with a northeast grain. The valleys are mostly underlain by limestones that do not resist erosion. Some geologists interpret the valleys as holes eroded through the thrust sheets to expose the rocks beneath the main thrust fault rocks, similar to those of the Vermont Valley and the Champlain-Hudson lowlands.

The Vermont Valley is a narrow, steep-walled, limestone valley from one to five miles wide that separates the Green Mountains from the Taconic Mountains between Brandon and the Massachusetts border, a distance of 85 miles. The valley is excavated in Cambrian and Ordovician strata that dip westward off the Precambrian core of the Green Mountains. The limestones are metamorphosed to marble. The Taconic klippe, with its basal thrust fault lies on these rocks on the eastern side of the valley. The distinctive topographic flavor of the Vermont Valley is most apparent in the widest part near its southern end at Bennington.

The Champlain lowland is a broad basin that separates the Green Mountains from the Adirondack Mountains of New York and holds Lake Champlain. Its western margin in New York is cut by numerous, steeply inclined block faults on which the Champlain side is dropped. The eastern side is cut by Taconian thrust faults that reactivated during the Acadian mountain-building event. The Cambrian and Ordovician strata of the valley floor lie nearly flat.

The Vermont side of the lowlands includes the Champlain Islands and a broad swath of land that rises gently to the foot of the Green Mountains. This is lake plain from Lake Champlain's much larger predecessor. The earlier Lake Vermont filled the basin while the northern escape for glacial meltwater through the St. Lawrence Valley was still blocked by ice, and the later Champlain Sea, an invasion of Atlantic waters, followed removal of the ice from the St. Lawrence. Shorelines of these earlier water bodies now rise northward because the land has risen since it was relieved of the weight of the ice. This is not a perfectly smooth surface; there are rock knobs here and there that were once islands. Also the surface has been modified by downcutting of streams since the close of the Ice Age. The bedrock surface is rather deeply mantled with glacial drift and lake deposits.

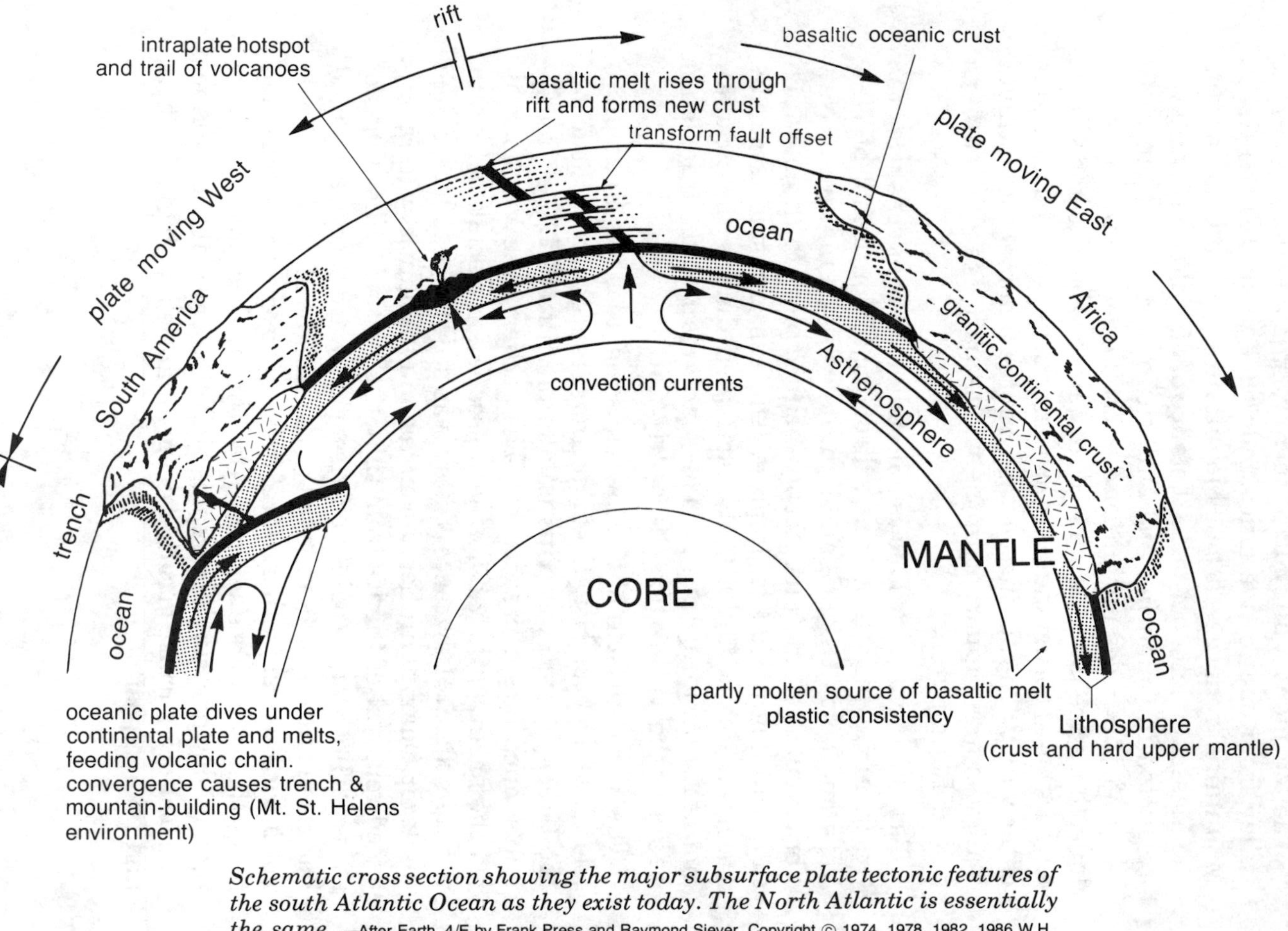

Schematic cross section showing the major subsurface plate tectonic features of the south Atlantic Ocean as they exist today. The North Atlantic is essentially the same. —After Earth, 4/E by Frank Press and Raymond Siever. Copyright © 1974, 1978, 1982, 1986 W.H. Freeman and Company.

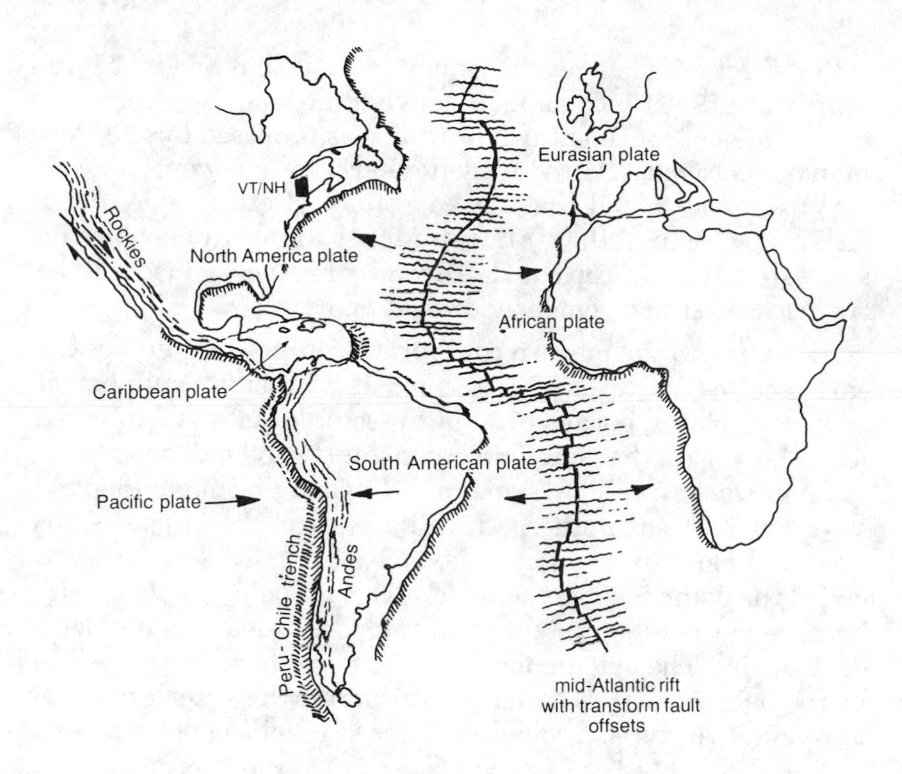

Major plate tectonic elements of the western hemisphere as they appear today. The Atlantic Ocean basin is still actively spreading as it has been for about 200 million years, shoving North and South America against the Pacific plate, producing an oceanic trench, mountain building, and volcanism.

Plate Tectonics: Vermont and New Hampshire and the Drifting of the Continents

The geology of Vermont and New Hampshire can only be understood in the context of much larger processes that shape the face of the earth. Let us now consider a broad geologic frame of reference called plate tectonics.

Plate tectonics is a revolutionary concept that had its beginning in the 1960s and has been snowballing since. It is rooted in the old idea of continental drift that was proposed by the German meteorologist Alfred Wegener early in the twentieth century. Wegener noted that the coastlines of either side of the Atlantic Ocean follow almost identical outlines, and he suggested that Europe and Africa were once joined to the Americas and had somehow drifted apart.

According to the modern concept, the outer thin shell of the earth, called the lithosphere, consists of a small number of enormous plates, perhaps 12, that resemble slabs of Arctic sea ice. Like sea ice, the plates move constantly and independently, causing some to collide, some to rift apart, and some to grind sideways past each other. Also like sea ice, the crustal slabs float and move over a denser layer beneath. This underlayer, called the asthenosphere, is solid rock that behaves plastically because of the high pressure and temperature associated with deep burial. The driving force for the movement is believed to be radioactive heat released within the earth's interior to produce giant convection cells like those in a boiling pot of water.

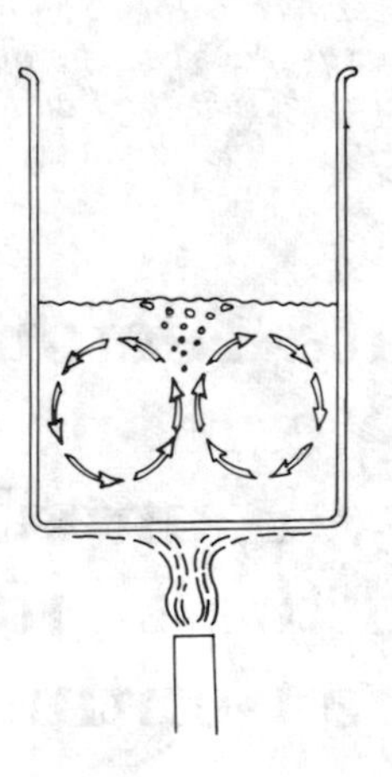

A familiar example of convection is seen when water is boiled over a Bunsen burner. Theoretically, the same mechanism causes convection in the asthenosphere, and tectonic plates ride the upper currents.

Cells that well up under the lithosphere and spread outwardly cause the plates to pull apart. Cells that converge and plunge back to the depths cause the overlying continental plates to come together and collide.

Most plates include segments of both thick continental and thin oceanic crust. For example, the North American plate encompasses North America, Greenland, half of Iceland, and half of the north Atlantic oceanic crust. The South American plate includes all of that continent and half of the south Atlantic. The African plate includes large portions of the ocean basins on all sides of the continent.

A prime example of a pull-apart plate boundary is the great submerged chain of volcanic mountains that traverses the length of the mid-Atlantic seafloor. This mid-Atlantic ridge has a deep rift valley along its crest. The mountains form by upwelling of lava from the asthenosphere through large fractures that open as the plates on either side of the rift pull apart. Each new lava flow becomes new oceanic crust; and each new rift splits that crust more or less down the middle. We can see that process at work in Iceland, one of the few places where rift lava erupts above sea level. Elsewhere, the oceanic crust is largely submerged.

Mount St. Helens and the other Cascade volcanoes of the Pacific Northwest coast lie parallel to a converging plate boundary where the denser basalt and mantle peridotite of the Pacific plate dive under the edge of lighter granitic rock of the North American continent. Parts of the descending crustal slab melt at depth to provide new magma for the volcanic eruptions.

The San Andreas fault is an outstanding example of the type of boundary where plates slide past each other. A series of jerks and pauses accompany the movement of the western California part of the plate, which is slowly drifting northward past the rest of the state and the continent. There is some basis in fact to the fears of Californians that a slice of their earthquake-wracked state may one day push off to sea. That "one day," however, will be millions of years long, because plate movement is only a few inches a year.

This shifting of crustal plates has been going on for a very long time, perhaps as much as half of Earth history. As a result, the positions, sizes, and shapes of the continents and ocean basins and all of their associated features are in a state of perpetual change. Ocean basins have repeatedly opened and closed. Mountain ranges arose along continental margins as thick piles of sediments and volcanic materials were squeezed during the closure, and when continents actually collided. Mountains along the sutures often achieved similar height and style to the modern fold mountain belts, such as the Urals or Himalayas.

Against this background, let us now consider the plate tectonics history of Vermont and New Hampshire. Exposed in southern Vermont within the core of the Green Mountains are the most ancient rocks of the two states. They record some of the earliest plate tectonics movements. The rocks are gneisses, quartzites, and marbles of Precambrian "Grenville" age whose date of metamorphism and folding lies between 1300-1100 million years ago. Grenville-age rocks also appear in the Berkshire Hills of Massachusetts, the Hudson Highlands, Manhattan Prong, and Adirondacks of New York, in scattered exposures in the Appalachian mountain chain to the southwest and in Newfoundland. None exist in New Hampshire. These ancient rocks extend beneath the cover of younger rock into the Grenville province of the Canadian Shield, a vast terrane of crystalline igneous and metamorphic rock that forms the nucleus of North America. West of the Grenville province the shield rocks are older and record earlier plate tectonic activity.

Hypothetical paleogeologic map of North America in late pre-Grenville time, as compared to the present outline of the continent. Note that the continental shelf, or platform, with its limestone deposits lies generally west of Vermont and New Hampshire.

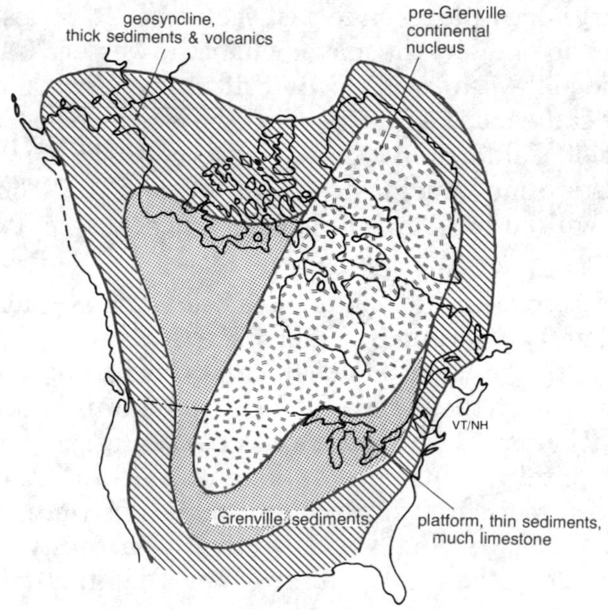

Taken together, the Grenville part of the shield defines an elongate, northeast-trending, several hundred mile-wide belt that extends from Labrador to Vermont and possibly as far as Mexico.

Many of the Grenvillian rocks began as sand, lime mud, and clay deposited in seas adjacent to a much smaller North American continent than exists today. This period of sedimentation spanned perhaps hundreds of millions of years when continental and oceanic crusts were locked together—neither rifting, colliding, nor sideslipping. The coastline then must have been a painfully desolate place: flat, devoid of both land plants and animal life. Mounds of blue-green algae called stromatolites poked their cabbage-like heads above shallow tidal flats. There were some coastal barrier islands and broad lagoons like those of the modern Gulf Coast, where salt and gypsum were deposited by evaporation. All the sediments gradually became tens of thousands of feet of solid rock.

The ocean basin began to close. Volcanic eruptions added their materials to the sedimentary pile. By about 1300 million years ago, the oceanic crust and upper mantle bordering the early continent slid into the Earth's interior, and the continents slowly came together, compressing the thick pile of Grenville sediments and volcanic rocks into a lofty folded mountain range that we'll call the ancestral Adirondack Mountains. Marginal strata from both colliding continents were crumpled, broken by faults, metamorphosed, and partly melted at great depth.

The continents seemed to enjoy a compatible relationship from the Grenville time of collision until their separation about 650 million years ago. By then, both had shed many miles of their rock covering so that the lofty bold mountains were reduced to a surface of little relief. The Grenville-age rocks exposed in Vermont today are the roots of these ancient mountain structures.

Crustal stretching some 650 million years ago may have caused the first movement on the numerous faults that are now so visible in the eastern and southern Adirondacks and Champlain and Mohawk lowlands. Continued rifting initiated not the present Atlantic Ocean but an earlier version called the proto-Atlantic. The shape and arrangement of the continents

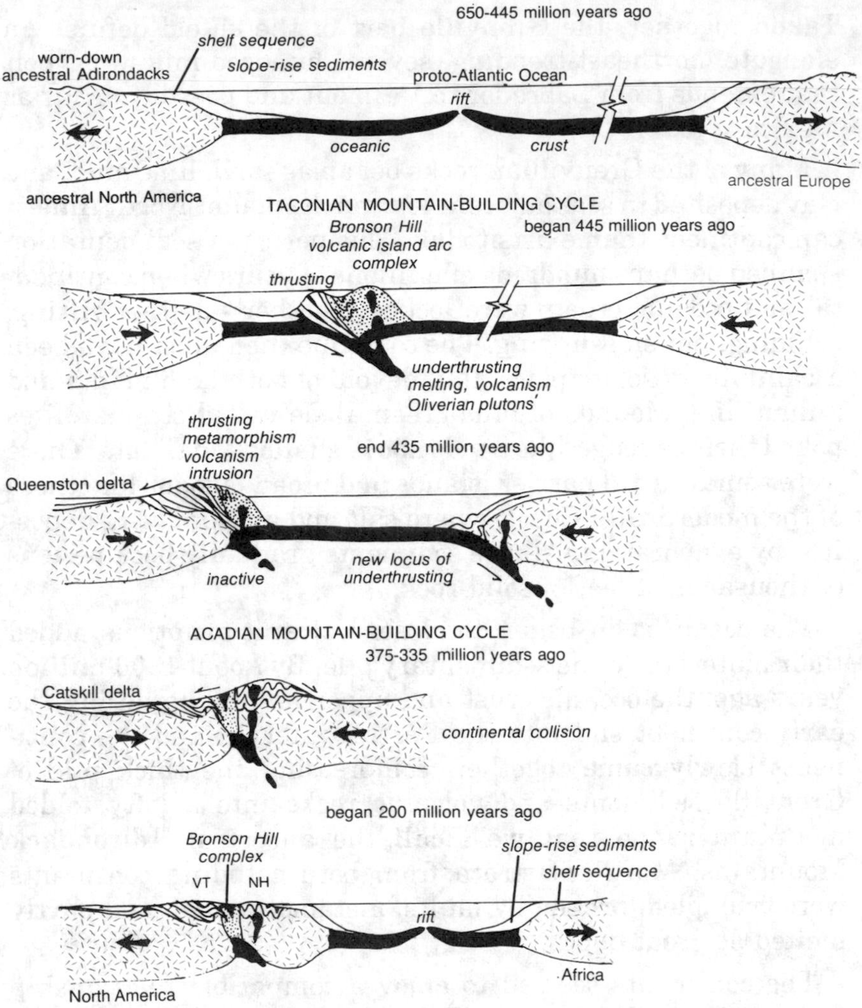

Hypothetical sequence of events leading to Taconian and Acadian mountain-building followed by opening of the modern Atlantic basin, beginning about 200 million years ago.

were quite different from today's familiar geography. The proto-Atlantic Ocean continued to widen until about 445 million years ago, as the mountains wore down and sediments piled up on the continental margin. The oceanic and continental crusts were tightly locked as parts of a single tectonic plate. Beginning about 520 million years ago, a shallow sea covered practically all of New England and New York. Early Paleozoic animals flourished as limestone, shales, and sandstones piled up. In Vermont, these constitute the "shelf sequence" of the Vermont Valley and Champlain lowlands.

A reversal of convection currents in the Earth's mantle began to close the proto-Atlantic Ocean about 445 million years ago. This was in middle Ordovician time, at the beginning of Taconian mountain-building. Initially, the compressional forces appear not to have broken the plate directly at the contact between thick continental and thin oceanic crust. Instead, the plate broke offshore, and the western segment of oceanic crust was shoved under the eastern segment. Some of the underthrust slab eventually melted; and the magma worked its way to the surface to form a chain of volcanic islands like those of Japan and the Phillipines. These ancient islands are called the Bronson Hill island arc complex; they consisted of both volcanic and intrusive igneous rocks. Volcanic materials and black muds were added to the sedimentary pile of the ocean floor and the edge of the continent.

Flexing of the continental margin by the underthrusting crustal slab in this early phase of the Taconian event either started up and down movement along the block faults within, east and south of the Adirondacks, or caused the blocks to shift again. Sediments and volcanic deposits were scraped from the underthrusting slab and piled upon the landward side of the islands. All deposits were intensely sheared and recrystallized at low-temperature but very high-pressure to form metamorphic rocks. The shearing extended into the dark mantle rock detaching innumerable slivers of it and shoving them up into the highly altered pile of deformed strata. These slivers have been converted to the dark green minerals serpentine and serpentine asbastos that now crop out in a continuous serpentine belt from north to south through central Vermont.

The island arc eventually "docked" or merged with the great mass of the continent when the intervening oceanic crust was completely consumed by underthrusting. Further crustal shortening followed: thrust faults increased in number and extended farther and farther inland, stacking great slices of continental shelf strata on top of each other to form the ancestral Taconic Mountains. In the late stages of compression the thrust faults extended into the crystalline Precambrian basement, shoving slices of these ancient rocks westward over younger Cambrian and Ordovician rocks that had originally been deposited on top of them. Rocks were folded at least as far inland as the Allegheny Plateau and Adirondack Mountains of New York. The famous Champlain thrust fault now exposed over the length of the Vermont side of the Champlain basin is the major fault at the base of the pile. Many rocks acquired a cleavage, consisting of closely-spaced fractures with very small displacements. Watch for the cleavage in the thin flat sheets of slate now so common in western Vermont.

Rocks buried deep in the core of the rising mountain range recrystallized during this period of thrust faulting and uplift. Sandstones were metamorphosed to quartzite; shales to slates, phyllites, schists, or gneisses; limestones to marble; and basalts to greenschists or amphibolites; the basement rocks were also modified.

By the time it was all over, the crust had been compressed an estimated 600 miles. It is difficult to say just how high the resulting mountains were, but it is certain that the metamorphism we see in the rocks happened under several miles of mountain overburden. An enormous amount of erosion has taken place since. Most of the eroded sediments spread westward, displacing a shallow sea to form an immense apron of many coalescing deltas collectively named the Queenston delta, after the late Ordovician Queenston formation of the Ontario lowlands in New York. So the Taconian event drastically changed the patterns of sedimentation in Ordovician time—from slow, eastward deposition on a gently sloping continental shelf to rapid, westward sedimentation from high fold and thrust fault mountains toward the continental interior.

Near the end of Taconian time, the delta plain was raised and eroded to produce the Taconic unconformity, a gap in the geologic record. Silurian and early Devonian sediments later accumulated on that erosional surface. In eastern New York the quartz pebble Shawangunk conglomerate rests on that erosion surface and supports the magnificent white cliffs and crags of the Shawangunk Mountains.

The Taconian event was followed by a period of plate tectonic calm between 435 and 375 million years ago. The oceanic slab attached to the continent no longer thrust under the edge of the collapsed island arc. The proto-Atlantic Ocean continued to close, but the compression was now relieved by a new rupture of the crust on the other side of the ocean; and the oceanic slab was shoved under the edge, of ancestral Africa and/or Europe. The Acadian mountain-building cycle, between 375 and 335 million years ago, marked the final closing of the proto-Atlantic basin and collision of continents. The backbone of the resulting, much loftier mountains lay east of the earlier Taconian Mountains.

Rocks as far west as the Adirondacks and Allegheny Plateau were newly deformed. The severe compression tilted the Shawangunk conglomerate backwards and shoved the entire Hudson Highland block of Precambrian rocks some distance northwestward. Thrust faults were reactivated. The old thrust slices crumpled into gigantic anticlines and synclines; many were ruptured by steeply-inclined reverse faults that further shortened the crust. The Precambrian rocks now exposed in the Green Mountains and the Athens and Chester domes of Vermont are pieces of continental shelf basement raised to a high level by Acadian faulting and folding, then later exposed by erosion. Ancient rocks and structures of the island arc complex are now exposed in the Bronson Hill anticline east of the Connecticut River. Principal rocks include the Ammonoosuc volcanics and Oliverian granites.

If this interpretation is correct, then much of New Hampshire must once have been a piece of Africa or Europe! This would

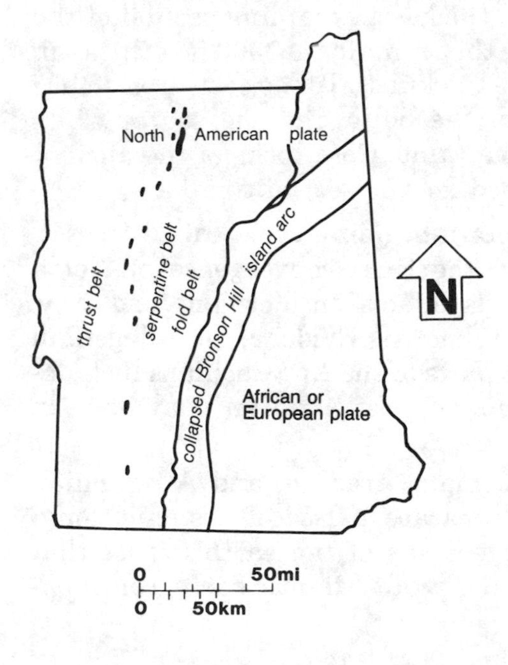

Major plate tectonic elements of Vermont and New Hampshire.

explain why the rocks of the two states are so different. New Hampshire would thus be "exotic terrain," continental crust formed elsewhere, then stuck to the edge of North America — and, of course, modified in the process. Acadian metamorphism was imprinted upon all rocks at least as far west as the Green Mountains, and it is possible that rocks even farther west were metamorphosed at low temperature during this time. The Acadian event is most notable for intrusion of igneous rocks that include the New Hampshire plutonic series — a pluton is simply a large body of magma that solidified at considerable depth beneath the surface. These igneous bodies are not restricted to New Hampshire. The famous Barre granite, used for tombstones throughout the United States, is one of many New Hampshire series plutons in Vermont.

Erosion of the Acadian Mountains provided raw material for another great delta. Like its Taconian counterpart, the Catskill delta was an enormous apron of sediments that spread westward to displace a shallow inland sea. These middle and late Devonian sediments completely buried the eroded stumps of the ancestral Taconic Mountains. The rocks formed from them are exposed today throughout the Allegheny Plateau principally in New York, Pennsylvania, Ohio and West Virginia. The present escarpment called the Catskill Mural Front rises high above the Hudson River; it is the erosionally truncated eastern edge and the thickest remaining section of the delta. The sedimentary pile that remains is 7500 feet thick, an estimated half of its original thickness. To the east, erosion has destroyed both the head of the delta and the source of its materials, the Acadian Mountains. Core rocks of the ancient Taconics have been exhumed for the second time.

The late Paleozoic mountain-building cycle called the Alleghenian orogeny marked the final convergence of North America and Africa. But its effects on Vermont and New Hampshire are virtually unknown. Evidence of Allegenian mountain-building in other parts of the Appalachians includes folding and metamorphism of post-Acadian coal beds in Pennsylvania.

Taken together, the Taconian, Acadian, and Alleghenian events can be considered separate pulses of a single, long-lasting convergence of large slabs of the earth's crust that terminated with closure of the proto-Atlantic basin, continental collision, and building of the eastern range of mountains that we now call the Appalachian chain. The timing and intensity of Acadian and Alleghenian mountain-building differed throughout the fold-belt, depending on when and where the continents actually touched when they collided. Only about half of the North American segment of the Appalachian chain is now exposed; the other half, a worn-down stretch about 180 miles wide, lies concealed under the sediments of the coastal plain and continental shelf.

At the end of the Paleozoic era a super continent called Pangaea existed. Pangaea contained all the ancestral continents fitted together like a jigsaw puzzle and surrounded by ocean. New mantle currents began to break up Pangaea during the Triassic period, about 200 million years ago, to form the modern continents and ocean basins. In the early stages, the crustal stretching produced the numerous block-fault basins of

Triassic fault block basins of the eastern United States.

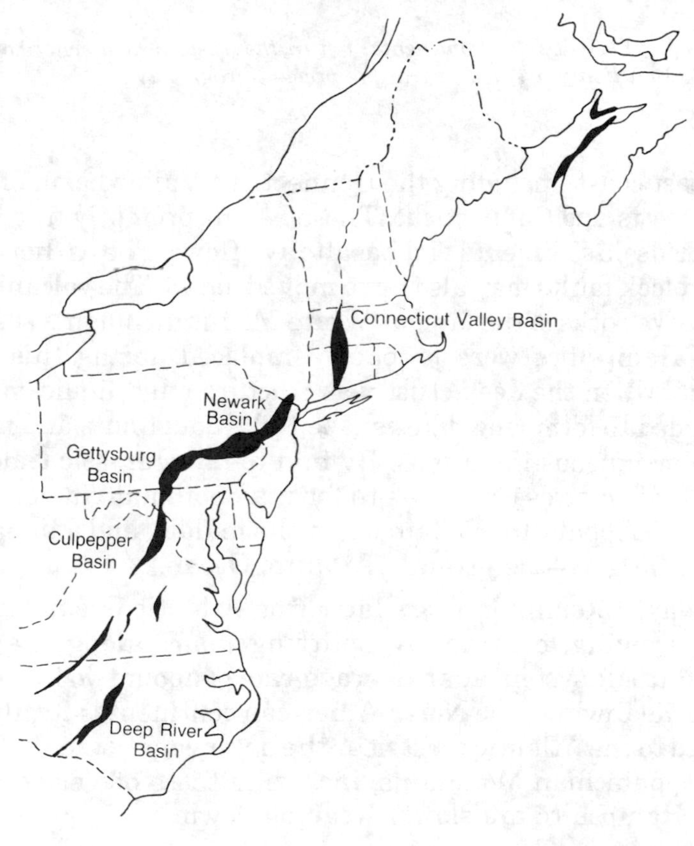

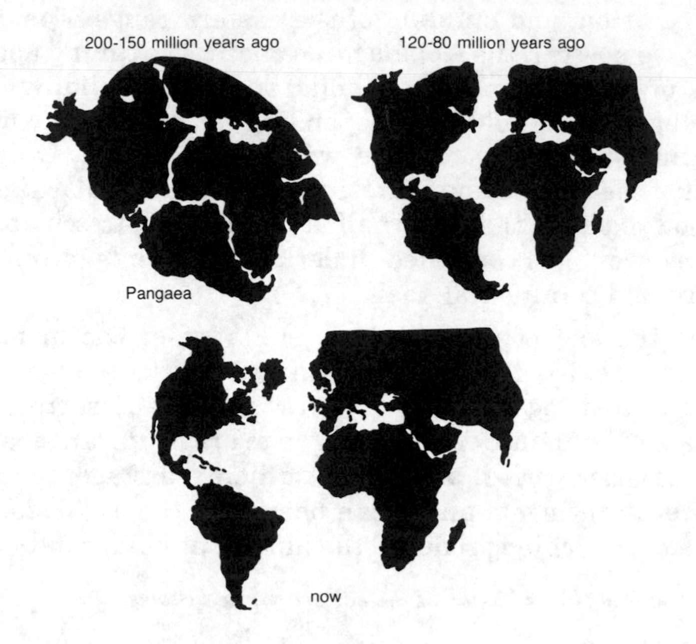

Breakup of Pangaea and continental rift in the western hemisphere over the past 200 million years, leading to the present geography.

the east coast, including the Connecticut Valley basin of Massachusetts and Connecticut. Those basins promptly filled with red Triassic sediments and basalt lava flows. The earlier Taconian block faults may also have moved again. The volcanic and intrusive rocks of the famous White Mountain magma series of New Hampshire were probably emplaced during this early period, when the deep crust was fractured and liquid magma ascended to form ring dikes shaped like doughnuts and nearly circular igneous intrusions. By mid-to-late Jurassic time, the smaller fractures connected to form a dominant rift zone that split the Appalachian chain down the middle, and a linear sea filled the gap—the juvenile Atlantic Ocean.

Today, 150 million years later, the Atlantic Ocean basin is still growing, continually splitting apart along the mid-Atlantic rift system at an average rate of about two inches per year. Meanwhile, the North American continent is firmly connected to the Atlantic crust near the outer edge of the shelf; and the Appalachian Mountains, including those of Vermont and New Hampshire are slowly wearing down.

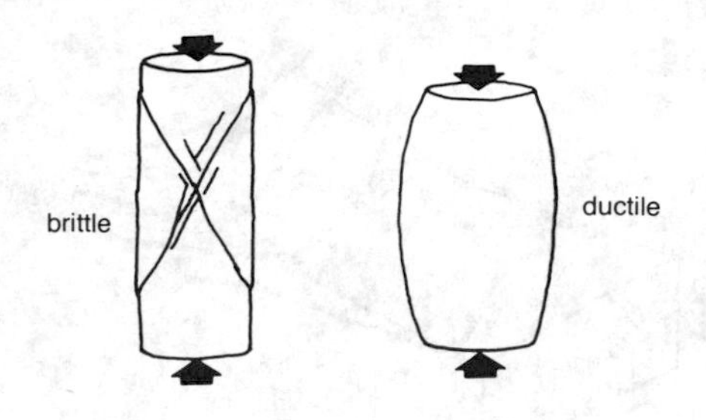

Brittle and ductile rock deformation are demonstrated in the laboratory by compressing two cylinders of the same rock. The one on the left faulted under conditions like those near the Earth's surface (unconfined, room temperature). The one on the right bulged under conditions like those at depth in the crust (high confining pressure high temperature).

Rock Structures: Folding, Faulting, and Jointing

Geologists call deformational features of rocks—structures, and their study—structural geology. Since rock structures are the clues to the large events of plate tectonics, let's look at some of the key structures you will see in Vermont and New Hampshire, and what they mean.

Thrust faults are the dominant structures of western Vermont. They indicate severe lateral compression and shortening of the crust at rather shallow depth. At shallow depth, rocks are brittle and will rupture under stress. Enormous thin slices can be pushed for many miles because they are rigid enough to transmit forces through them. The accompanying drawing schematically illustrates the concept using a number of Geological Society of America bulletins arranged in "imbricated" fashion as shown, and at first spread out widely. These are the thrust slices, the lowest of which (the Champlain slice) leans against the New York foreland, here represented by the book *Roadside Geology of New York*. The right edge of the

Westward stacking of thrust slices in Vermont against the New York "foreland" is demonstrated with several issues of the Geological Society of America Bulletin. The bulletins slide past each other as do large discrete thrust slices.

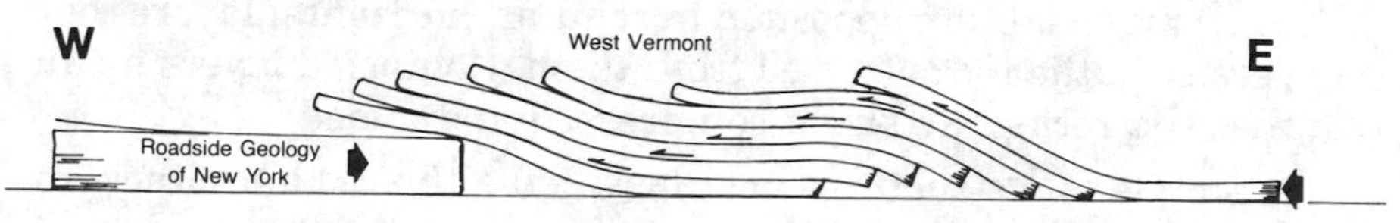

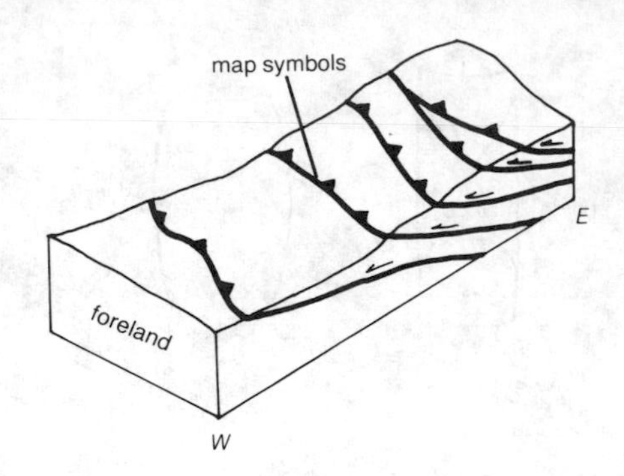

Thrust slices are stacked on top of each other in western Vermont more or less as shown in this schematic block diagram.

easternmost slice is now held flat on the table and shoved slowly westward while holding the foreland fixed. The result is that the slices thrust over each other and pile up westward. The lateral force is transmitted through them because they are sufficiently rigid, even though they bend easily. As they bend, the pages also shift minutely, similar to the internal, microscopic shearing of the real rock that is largely responsible for its strongly layered character. As might be expected, rocks nearest the thrust faults are normally the most intensely sheared; this, in fact, is often the best evidence that there is indeed a thrust fault nearby.

The slight metamorphism of most of the thrust slice rocks also points to a shallow origin for, in general, metamorphic grade increases with depth of burial. Most of the rocks are slates, phyllites, and low-grade schists, quartzites, and marbles.

Unfortunately, thrust faults are inconspicuous in outcrop and roadcut. Regional geologic mapping or topographic expression provide the primary evidence of their existence. There are well-defined scarps, for example, on the leading edge of the Champlain slice at Mount Philo, Buck Mountain, and Snake Mountain held up by resistant Monkton quartzite, and on the Hinesburg thrust supported by the Cheshire quartzite.

Rock folding is a much more visible and widespread form of deformation in the two-state region. Most sediments and volcanic materials are deposited in regular, horizontal layers that persist in the rocks formed from them. Contorted layers mean that the rocks have been compressed and folded.

Simple fold structures are classified either as anticlines, in

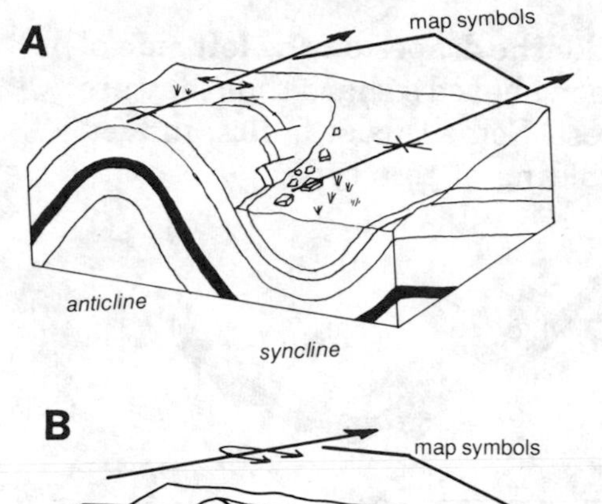

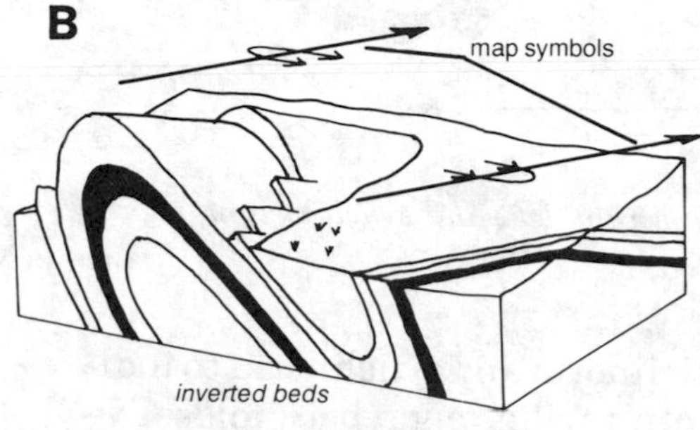

Schematic block diagrams showing (A) upright and (B) overturned folds with appropriate map symbols.

which the layers are upfolded, or synclines, in which they are downfolded. Folds seldom exist alone; but occur in multiples of alternating anticlines and synclines. On a regional scale, they resemble the folds in a drape. This analogy also serves to illustrate how the folds form and how they shorten the crust. When the drapes are closed, they are only slightly folded. As they are opened by pulling the drawstring, the folds compress, and the same material is crowded into a narrow space at the side of the window. In Vermont and New Hampshire, the metamorphic rocks show a north to south corrugation because

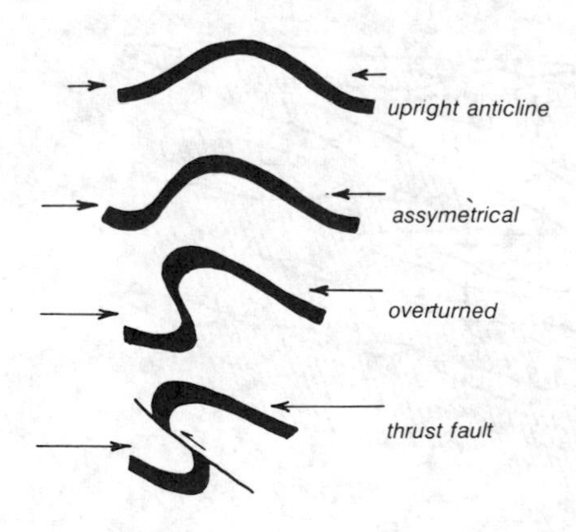

Schematic sequence illustrating how some thrust faults develop from folds.

they were shoved westward—like the drapes on the left side of the window. Many folds have been shoved so hard that they are overturned; they lean to the west. Some thrust faults, in fact, may be outgrowths of severe folding of this type.

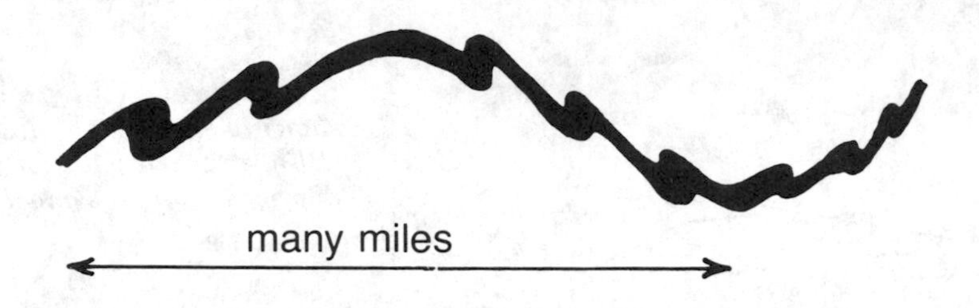

Schematic cross section of an anticlinorium, left, and synclinorium, right.

Geologists use the terms anticlinoria and synclinoria to indicate large-scale folds that contain smaller piggyback folds. I've avoided these terms altogether to keep things as simple as possible. It is worth noting, however, that the Hinesburg, Middlesbury, and other large synclines are really synclinoria, and the Strafford-Willoughby arch, Bronson Hill, and other large anticlines are really anticlinoria.

Fold character varies considerably and often reveals a great deal about depth of burial at the time of deformation. Almost all rocks at the surface are brittle and tend to break rather than

A fluid, or very ductile, fold looks something like this with pinched-out layers on the limbs; this style usually means deep burial at the time of folding.

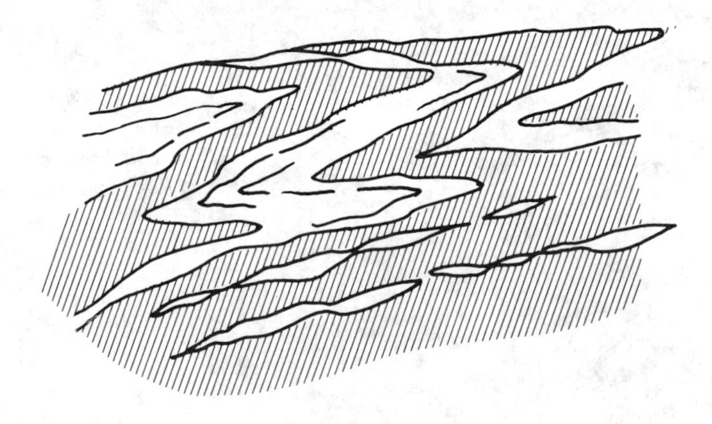

bend, while rocks buried deeply are ductile or taffy-like and will bend, even flow, easily. The slates of western Vermont, for example, exhibit a brittle folding style indicative of shallow depths, consistent with their low grade of metamorphism. Folded layers within the slate are not so much flexed as offset by closely-spaced cleavage planes, as shown in the accompanying schematic drawing sequence. On the other hand, many folds in the schists and gneisses display more fluid patterns consistent with conditions of deep burial and the higher grade metamorphism recorded in the minerals that compose them. Pinch-and-swell of some layers generally reflects the ductile nature of the material during folding.

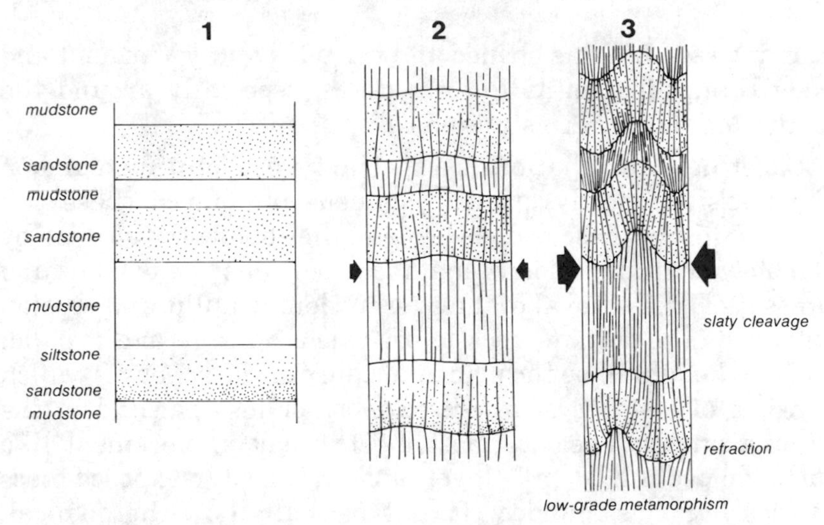

Formation of axial plane slaty cleavage by lateral compression accommodates crustal shortening and produces folding. Note the change in cleavage direction, or refraction, across boundaries between layers of different composition.

In addition to the above "push" structures, western Vermont, including the Champlain Islands, reveals many pull-apart structures called block faults. As with thrust faults, regional mapping is the primary way to find them.

Joints are simple fractures visible in almost any outcrop or roadcut of the two-state region. Unlike faults, the rocks on either side do not slide past each other—the fractures just open up as the rock contracts, although the opening is typically microscopic. Joints typically occur in sets of numerous parallel joints, and a single outcrop may exhibit two or more criss-

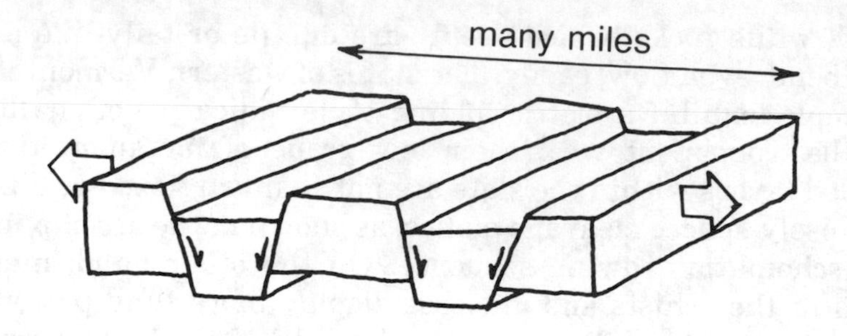

Schematic illustration of block faulting in response to crustal stretching.

crossing sets. Magma opened and intruded many Vermont and New Hampshire joints to form dikes, especially around the White Mountain series plutons.

Sheeting, or exfoliation, is a special type of jointing found on the massive, mostly granitic, igneous plutons of these two states. Most of these bodies formed miles below the surface by the slow crystallization of magma under enormous confining pressure. The removal of the overburden by uplift and erosion permitted the plutonic rock to breathe a big sigh and expand. This is thought to be the principal cause of the sheeting, which consists of a series of fractures more or less parallel to the ground surface. The outermost sheets break off piecemeal, like an onion peeling layer by layer; and, in time, the exposed mass of rock becomes rounded. This is basically how the domical, barren summits of the White Mountains were sculptured.

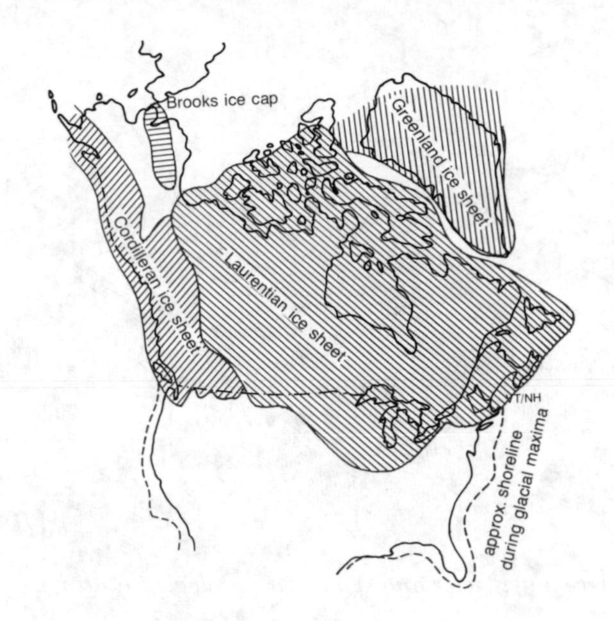

Approximate maximum extent of the major ice sheets in North America during the Ice Age. At climax, so much water was locked in the ice that sea level was as much as 300 feet lower and the coastline far out from its present position, as shown by the dashed line.

The Ice Age—Vermont and New Hampshire Glaciated

To better illuminate the glacial geology of Vermont and New Hampshire, let's review some of the history of the Pleistocene, or Ice Age, epoch of geologic time. The Ice Age in North America, and in Europe, began earlier than two million years ago, and large-scale continental glaciation ended approximately 6000 years ago. At least four major glacial advances and retreats triggered by fluctuations in the Pleistocene climate marked this long episode. In terms of geological time, the advances were brief and the interglacial periods long. There is erosional and depositional evidence for all four major advances in other states, but Vermont and New Hampshire contain only features of the latest or Wisconsin glaciation. This latest surge of ice apparently swept away or covered the clues to earlier advances.

Wisconsin glaciation climaxed 18,000-20,000 years ago

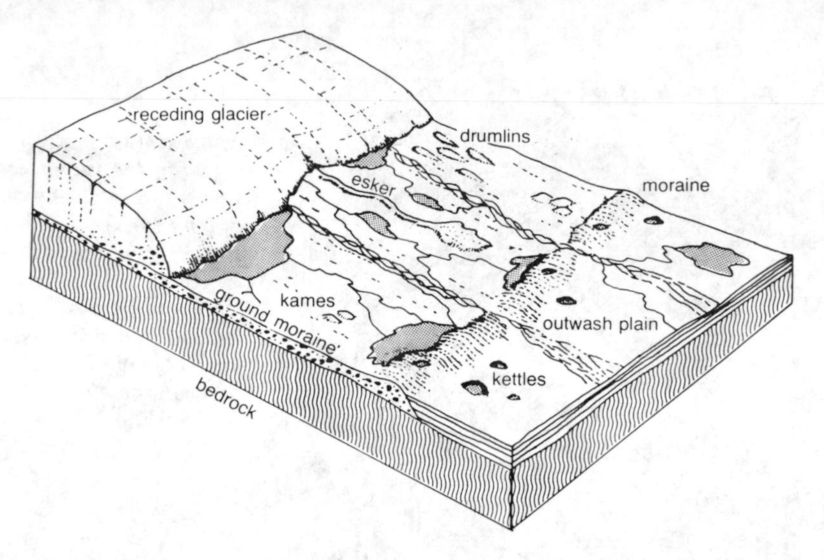

Schematic diagram of a receding glacier and some of the geologic features it leaves behind

when ice covered most of Canada, as well as the sites of Seattle, Chicago, and New York City. All of New England lay buried under ice that was more than a mile thick in places. The glacier overrode all of the mountains of Vermont and New Hampshire, softening the topography and stripping off the soils and earlier glacial deposits. The ice scraped the north sides of the mountains and plucked their southern slopes to give the peaks a lopsided profile with gentle north slopes and south-facing cliffs. Thick tongues of ice coursed through the valleys that were more or less in line with the ice flow and gouged them deeper and wider. At climax, the Wisconsin ice sheet reached southward to the south fluke of Long Island, Block Island, Martha's Vineyard, and Nantucket Island. The ice terminus remained fixed long enough to dump much ice-bound sediment and form a ridge of till called the Ronkonkoma terminal moraine. The Harbor Hill recessional moraine marks a later stand farther north on the north fluke of Long Island, southern Rhode Island, and Cape Cod.

Glacial erosion is more evident than deposition in Vermont and New Hampshire. Neither state contains well-defined moraines marking long stands of the glacier front, nothing like the Ronkonkoma and Harbor Hill moraines.

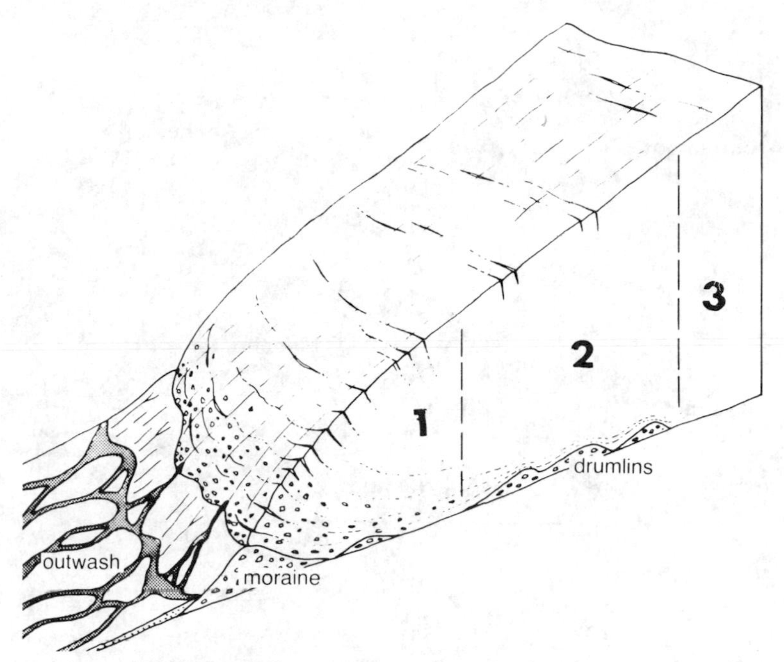

Conditions for the formation of drumlins. In zone 1, the weight of the ice is low and sediments are simply melted out and dumped in place to form moraine. In zone 2, the weight of the ice is just right to cause plastic flow in till and shape it into drumlins. The excessive weight of the ice in zone 3 either causes the till to be bulldozed away or smeared along the ground.
Courtesy Kendall/Hunt Publishing Co.

Drumlins, those peculiar ice-molded, streamlined hills of till, exist in abundance in southern New Hampshire, but not in very perfect form. Thin ground moraine laid down under the ice or while the glacier melted is widespread, clearly evident in many fields littered with erratic boulders. Early settlers made the famous New Hampshire stone fences of these boulders as they cleared the fields for farming. Outwash sediments, transported and deposited by glacial meltwater streams, are largely confined to valleys. Some valleys, particularly those of the Connecticut and Merrimack rivers, contain thick deposits of sediments laid down in lakes that formed along the edge of the ice. Such deposits include clays composed of fine rock "flour," produced by grinding of ice-bound rocks against each other and against the bedrock. This pulverized material is what makes glacial streams milky. It is so fine that it will only settle out in the quiet water of a lake or sea.

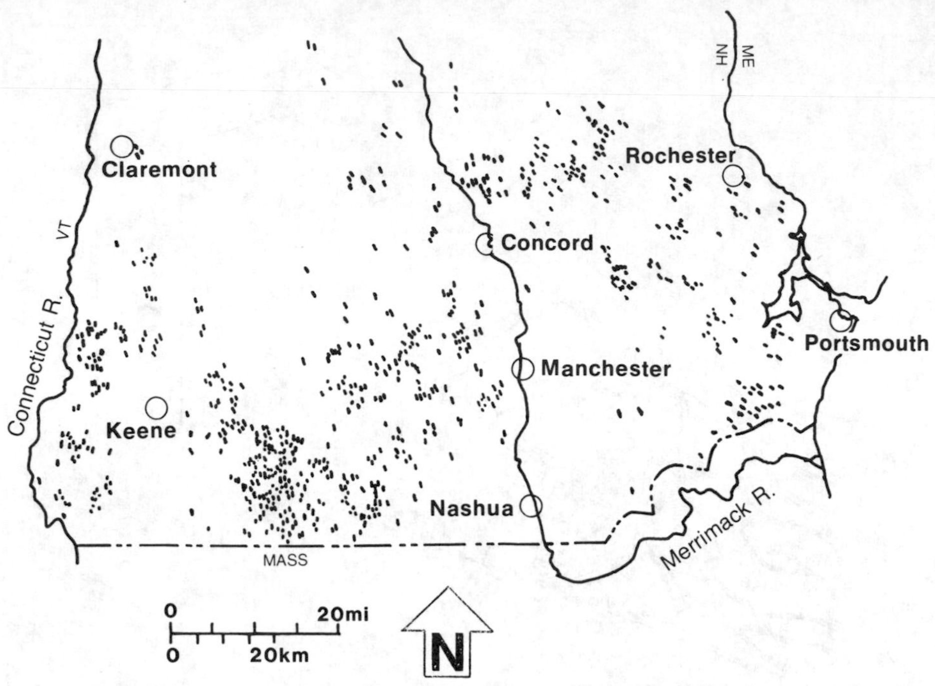

Drumlins of southern New Hampshire, part of a larger field that extends southward into Massachusetts. Many in New Hampshire appear to be bedrock-cored.

The clays are uniquely deposited in annual layers called varves that record the summers and winters of the ice age like tree rings record the seasons of the tree's years. It is possible to count the years of sediment accumulation, sometimes to determine when a particular site was first freed of ice.

Other common glacial lake deposits in Vermont and New Hampshire are the kame terraces that mark the sides of many valleys. These are actually strip deltas deposited by streams flowing along the edges of valley ice tongues while the glacier melted. Such streams emptied into lakes along the ice margin where they dumped their sediment load. Their points of sediment discharge shifted with the receding ice front, producing a continuous terrace.

Kame terraces and other outwash deposits tend to be sandy, stratified, at least moderately size-sorted, and free of clay and silt. By contrast, till, deposited directly from melting ice, commonly contains so much clay and silt, as well as sand, gravel, and even huge boulders, that it is sometimes referred to as "boulder clay."

Not all of the conspicuous terraces in Vermont and New Hampshire valleys are kames. Most are remnants of valley-fill sediments that escaped erosion as the streams began to erode after the land, freed of its burden of glacial ice, floated upward. It is this process that exposed the many varved clays in the Connecticut and Merrimack valleys.

For some reason, eskers are uncommon in these two states, although they are exceptionally large and abundant in adjacent Maine, as are drumlins. An esker is a ridge of sediment deposited along the course of a subglacial, tunneling stream. They typically trend more or less parallel to the ice flow direction. Like kame terraces, eskers form as the ice melts. They may appear, at least from the air, like "rivers of sand," complete with meanders and branching tributaries. The Passumpsic Valley esker near St. Johnsbury, Vermont, is the largest and best-developed one in these two states; it continues virtually unbroken for 24 miles.

Kettles are pits formed where blocks of ice left behind by the receding ice front were partly or wholly buried in outwash sediments before they melted. Filled with water, the pits became kettle lakes. They are common in moraine areas and in the outwash deposits near them.

In the early stages of the Wisconsin glacial episode, before the ice sheet covered the land completely, small alpine glaciers coursed down many of the valleys high upon the flanks of the highest peaks of the Green and White mountains. They lasted long enough to carve deep, bowl-shaped rock amphitheaters, or cirques, into the mountain sides. The famous Tuckerman and Huntington ravines on the east side of Mount Washington are spectacular testimony to this event. It is possible that the alpine glaciers also reformed in late-Wisconsin time, when the climate at high altitude was still glacial enough to maintain them, although the global climate no longer permitted the continental ice sheet to reach this far south. There are even more profound cirques on Mount Katahdin in Maine; other, less well-developed ones exist in the Adirondack and Catskill mountains of New York.

As the Pleistocene epoch drew to a close, numerous temporary lakes formed between ice and high ground. They were the receiving basins for the copious sediments shed from the melting ice and the barren land nearby. The spillover outlets for these lakes, which controlled water levels, changed frequently as the ice receded. Each time a new spillway was uncovered at a level lower than the existing one, the lake level dropped, and its shoreline changed accordingly. Glacial geologists have given numerous names to these lakes. Lake Vermont, for example, is the name given to the very large one that filled the Champlain basin to levels far above that of modern Lake Champlain while the northern end of the basin and the St. Lawrence River were still ice-blocked. Lake Vermont's

Major glacier lakes of Vermont and New Hampshire and marine submersion of New Hampshire seacoast. Outlines are approximate, based on Stewart and MacClintock.

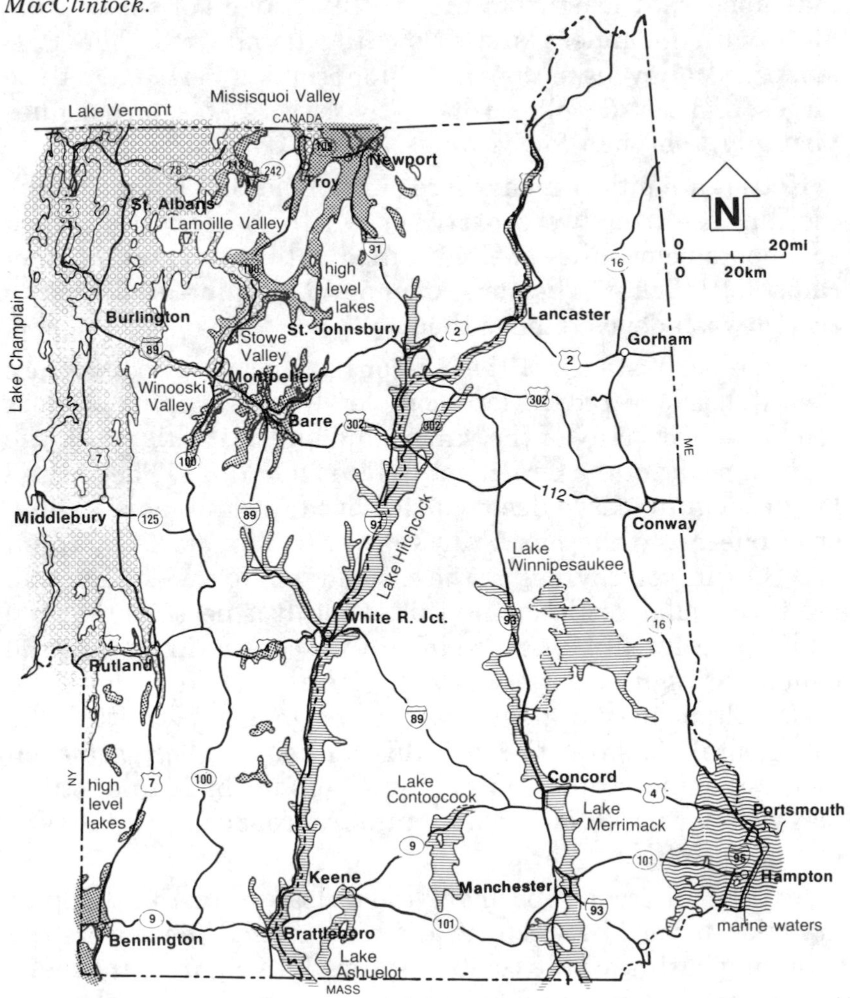

shorelines reached 10-20 miles from the present shores to the foothills of the Green Mountains and fingers even crossed the mountains via the Missisquoi, Lamoille, and Winooski valleys. The Connecticut Valley Lake, also called Lake Hitchcock, achieved a maximum length of well over 200 miles, stretching like a snake from Middletown, Connecticut, to the Canadian border. It was dammed at its southern end, as the Finger Lakes of New York are presently dammed, by glacial moraine, which was eventually breached by outflow. Actually, the Connecticut Valley Lake probably consisted of numerous long narrow lakes separated by stagnant ice blocks. Either way, the conditions were nearly ideal for the deposition of varved clays that have helped geologists to chart ice recession.

Where proglacial lakes stood at a certain level for a long time, wave action carved cliffs and benches and formed river deltas, dunes, and other features like those along modern lakeshores. Lakes Vermont and Hitchcock shore plains have been most useful in charting the pace of uplift of the land since the enormous weight of the ice was removed. By tracing originally horizontal shorelines from south to north, it has been determined that the northern Champlain lowland surface has risen more than 500 feet. The amount of rebound diminishes southward because the ice was thinner there, the land less depressed under its weight.

As the ice receded into Canada, the St. Lawrence Valley opened to the sea, and Lake Vermont drained to a low level. Eventually, Atlantic waters invaded the St. Lawrence-Champlain lowlands, forming the Champlain Sea. The shore and bottom sediments of this embayment may often be recognized by the marine fossils they contain. Glacial rebound gradually diminished the inland reach of sea water to its present extent.

The manner of recession of the last ice sheet from New England is a subject of great controversy, with opinion divided between "retreaters" and "downwasters." Most glacial geologists now concede to both processes. The retreaters envision deglaciation as a gradual shrinking of an active, flowing ice sheet. The ice margin at any one moment of geologic time was where the rate of climate-controlled ice wastage exactly equaled the rate of supply of new ice moving southward in the ice sheet. As the climate warmed, the margin shifted north-

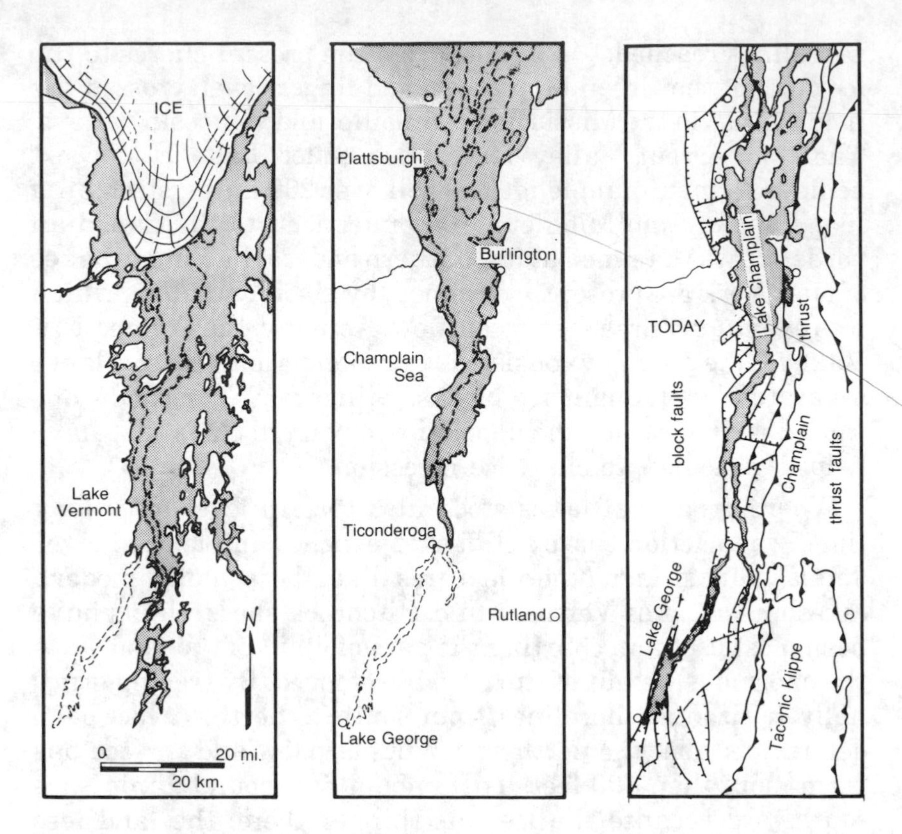

Three maps of lake stages in Champlain basin. —Courtesy Kendall / Hunt Publishing Co.

ward to a new line of balance between supply and demand. In a sense, the ice sheet pulled in its borders while it nevertheless continued its southward flow. The downwasters, on the other hand, stress the importance of stagnation, the de-activation of glacial flow and fragmentation of the ice sheet as it thinned over large areas of uneven terrain. The latter process would result in numerous small meltwater lakes temporarily dammed by ice blockages. The patchwork of deltaic sand and gravel deposits and lake bottom sediments in New Hampshire seems to support this interpretation. Most of these lakes were not large enough and did not last long enough to form well-developed shoreline features like those of Lake Vermont.

Glacially polished and scratched bedrock exposures appear in almost all parts of the states. Very large boulders pressed hard against bedrock gouged deep, wide furrows. Some rocks

Glacial erratic about 11 miles south of Conway.

advanced in small jumps and left rows of crescent-shaped fractures in the bedrock surface, chatter marks. All of these features are best preserved in the more durable rocks, like quartzite; in limestones, schists, gneisses, and granites, they are easily destroyed by exposure to the weather. Glacier-bound boulders also grind against each other, creating polished and scratched surfaces that distinguish them from rocks that were carried by streams. All glacially transported boulders may be referred to as erratics.

Sheepbacks are elongate knobs of bedrock, the smaller counterparts of the glacially sculptured summits of high mountains. They are elongated in the direction of ice flow, with gentle upstream sides that were scraped smooth by oncoming ice and steep, craggy, downstream ends from which rock was plucked and carried away by the ice.

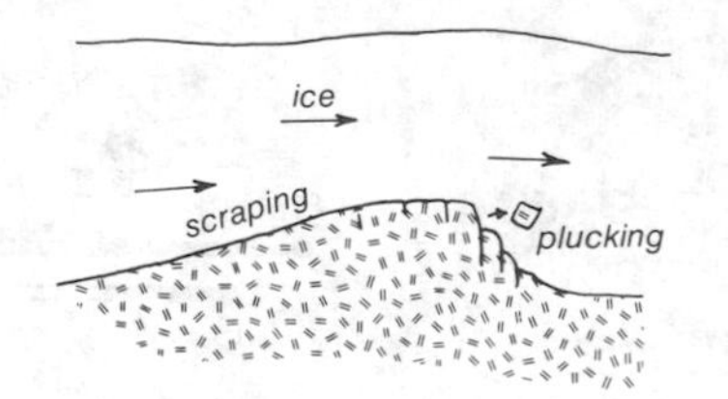

Formation of "sheepback" by passage of glacial ice over bedrock knob. The same process shaped many Vermont and New Hampshire mountain summits.

89

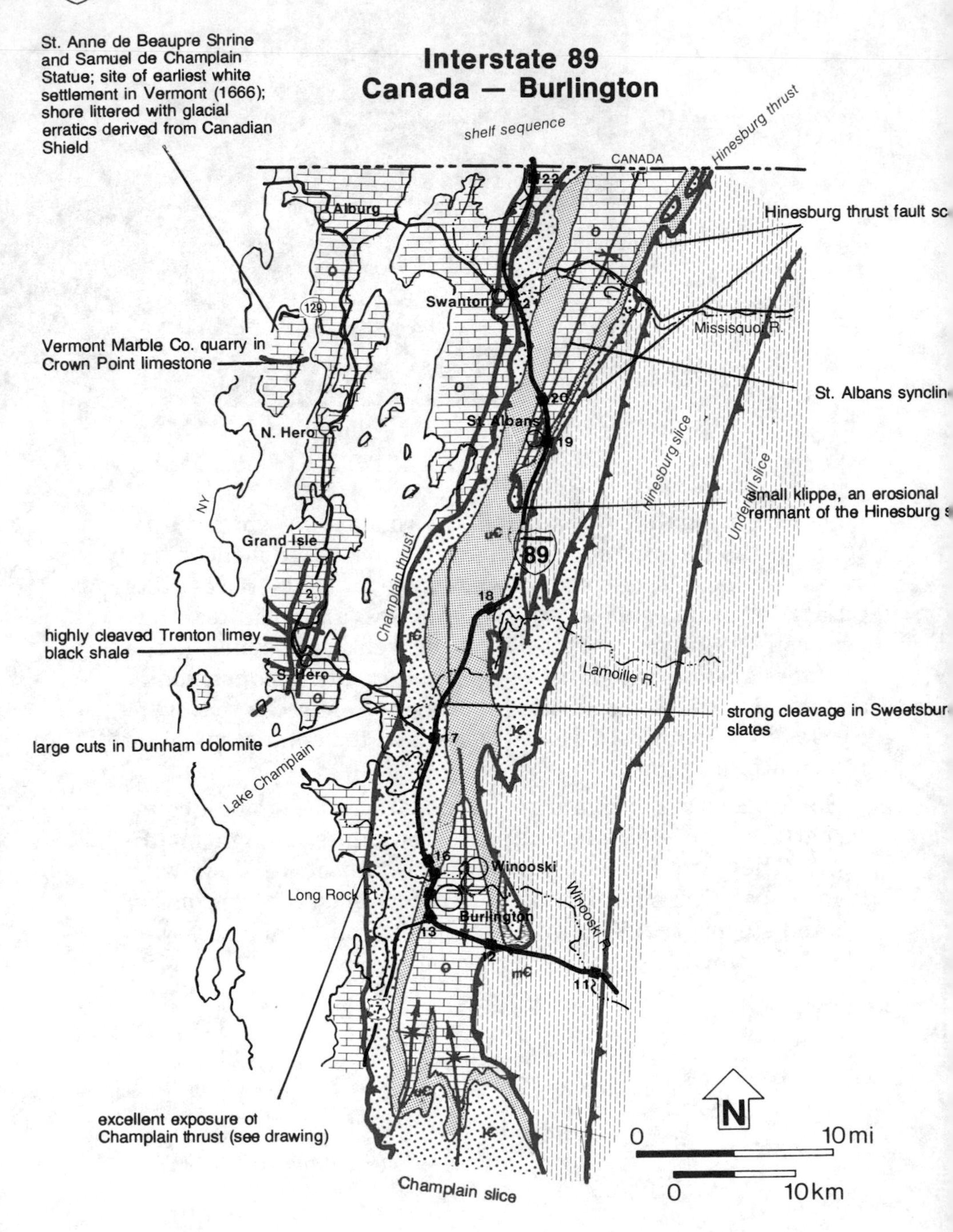
Interstate 89
Canada — Burlington
St. Anne de Beaupre Shrine and Samuel de Champlain Statue; site of earliest white settlement in Vermont (1666); shore littered with glacial erratics derived from Canadian Shield
shelf sequence
Hinesburg thrust
CANADA
VT
Hinesburg thrust fault sc
Alburg
Swanton
Missisquoi R.
Vermont Marble Co. quarry in Crown Point limestone
St. Albans syncline
St. Albans
N. Hero
NY
Hinesburg slice
Underhill slice
small klippe, an erosional remnant of the Hinesburg s
Grand Isle
Champlain thrust
highly cleaved Trenton limey black shale
Lamoille R.
S. Hero
strong cleavage in Sweetsburg slates
large cuts in Dunham dolomite
Lake Champlain
Winooski
Long Rock Pt.
Burlington
Winooski R.
excellent exposure of Champlain thrust (see drawing)
Champlain slice
N
0
10 mi
0
10 km

Interstate 89: Canadian Border—Burlington, Vt.

40 mi. / 64 km.

Vietnam Veterans Memorial Highway

Between the Canadian border and Swanton (6 miles) Interstate 89 follows the Champlain thrust, the westernmost great thrust fault of Vermont that continues southward nearly the full length of Lake Champlain. Here, there is not much of a fault scarp, but it becomes locally prominent south of Burlington, where it places lower Cambrian Monkton quartzite over weaker rocks. Visible 7 miles to the east is the much higher scarp of the Hinesburg thrust fault, capped by lower Cambrian Cheshire quartzite. The road crosses from the west to east side of the Champlain thrust at exit 21, just before passing over Rock River. Roadcuts and rocks exposed in the riverbanks are shattered near the fault.

All rocks on this route were metamorphosed at low temperature. They are really marbles, instead of limestone or dolomite, actually slate instead of shale, despite the wide usage of sedimentary rock names. The deformation and metamorphism are probably the combined result of Taconian and Acadian mountain-building. As noted in the plate tectonics chapter, the Acadian Mountains were centered to the east of the earlier Taconian Mountains, so deformation and metamorphism associated with their formation diminish westward.

Between Swanton and St. Albans (6 miles), you cross diagonally over the St. Albans syncline, a very large downfold with numerous superimposed small folds in Cambrian and Ordovician strata of the lowland belt between the Champlain and the Hinesburg thrusts. The Hinesburg scarp is profiled to the south by St. Albans. Between St. Albans and Burlington (23 miles) are many lovely views over Lake Champlain and New York's Adirondack Mountains.

Topographically, the Champlain lowlands extend to the foot of the Green Mountains, including the belt floored by the Champlain slice. From a structural standpoint, the Champlain thrust fault marks the eastern limit of the lowlands, which are underlain by rocks that are still in their original location and relatively mildly deformed. Everything east of that is out of place and more severely deformed by thrust faulting.

The depression the lake floods was created both by deformation and erosion. It is, first of all, a structural basin between the Precambrian massif of the Adirondacks and the thrust faults of western Vermont. Cambrian and Ordovician strata are gently downfolded as well as precipitously downfaulted along numerous block faults. The latter, which are most conspicuous on the New York side of the lake, are steeply-inclined faults distinguished by vertical movement. Lake George of New York is in an excellent example of a fault block basin.

Streams and glaciers hollowed the structural basin deeper. Ice invaded the lowland on at least 4 separate occasions during the Ice Age. The alignment of the valley with the direction of ice flow channeled the ice each time, forming a Champlain lobe that gouged the depression deeper and wider as it reduced the existing hills.

The muted landscape of the Champlain lowlands is also partly the result of extensive sedimentation. In the closing stages of the Ice Age, as the margin of the Champlain lobe gradually receded northward, glacial Lake Vermont filled the basin to levels much higher than the present lake, reaching inland to the Green Mountains. Lake Vermont later drained to the sea when melting ice opened the Richelieu and St. Lawrence valleys. Still later, after a lengthy period of erosion, Atlantic seawater invaded the basin to form the Champlain Sea. The net

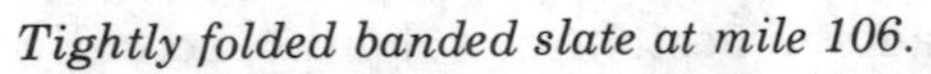

Tightly folded banded slate at mile 106.

Strongly developed cleavage in Sweetsburg slate, 4 miles south of exit 18. Folding here is largely accomplished by small offsets along cleavage rather than flexing of the beds.

result is that much of the existing bedrock relief is mantled with flat-lying lake and marine sediments, mostly clay and silt.

At St. Albans, again about 4 miles south, you pass just east of two small rocky hills that appear to be erosionally isolated remnants, or klippen, of the Hinesburg thrust-fault slice. Early Cambrian Cheshire quartzite caps the hills. These rocks are out of place, shoved westward and left sitting on top of younger late Cambrian Sweetsburg slates, to which they bear no genetic relationship. Small klippen are very common in western Vermont; the southwestern part of the state is dominated by the enormous Taconic klippe, which continues southward in New York to the Hudson highlands as the caprock of the Taconic Mountains.

Near exit 18 to Georgia Center and Fairfax, just north of the Lamoille River bridge, are several large roadcuts with wild folding. One in particular, 4 miles south of the exit, shows well-developed cleavage in dark Sweetsburg slates, delineated by offsets in sharply-defined lamellar bedding. This is axial-plane cleavage that transects the bedding and offsets it in innumerable tiny steps that create the fold pattern defined by the bedding. The fold forms without much actual bending of the beds. This exposure lies very close to the axis, or line of sharpest folding, of the Hinesburg syncline, another large downfold in the Champlain slice. The contrasting fold type, called a flexural fold, is visible on this route in some of the thicker-bedded marbles.

Quarry in Monkton quartzite at Winooski. With blasting, the thinly-bedded, brittle quartzite breaks up into small, jagged fragments that bind well in concrete.

South of exit 17, the route crosses Malletts Creek near its mouth at Malletts Bay. Here, the stream cuts through the early Cambrian Monkton quartzite, and the road follows this same unit for about 5 miles to Winooski. The Monkton quartzite is a distinctively red- to buff-colored, thin-bedded rock that also contains relatively thick layers of marble. A large quarry at Winooski, next to exit 16, produces quartzite for high quality crushed aggregate for concrete. A total of 708,000 tons came from this pit in 1984. The rock is hard and inert, and it fractures into angular fragments that bind well in

Champlain thrust exposed on Long Rock Point near Burlington. The middle Ordovician shale below the thrust is highly sheared and contorted; overlying massive rock is early Cambrian Dunham dolomite.

concrete. The rock also breaks up well with blasting, minimizing the need for crushing.

Between exits 15 and 14 to Burlington (2 miles), Interstate 89 crosses the Winooski River over whitish Ordovician marbles in the Champlain slice. The thrust fault itself is exceptionally well-exposed just west of the city on Long Rock Point slightly above Lake Champlain water level. Here, it places the massively bedded, buff to gray, early Cambrian Dunham dolomite over black shales of the younger middle Ordovician Iberville formation. The thrust fault is sharp, marked by a thin, discontinuous zone of crushing that contains angular dolostone fragments embedded in a highly contorted matrix of shale. Slivers of limestone, some several feet thick, that may have been ripped from the early Ordovician Beekmantown group that is transected at depth by the fault farther east, have been dragged westward in the fault zone. The fault surface dips, or slopes, southeastward at an average 10 degrees.

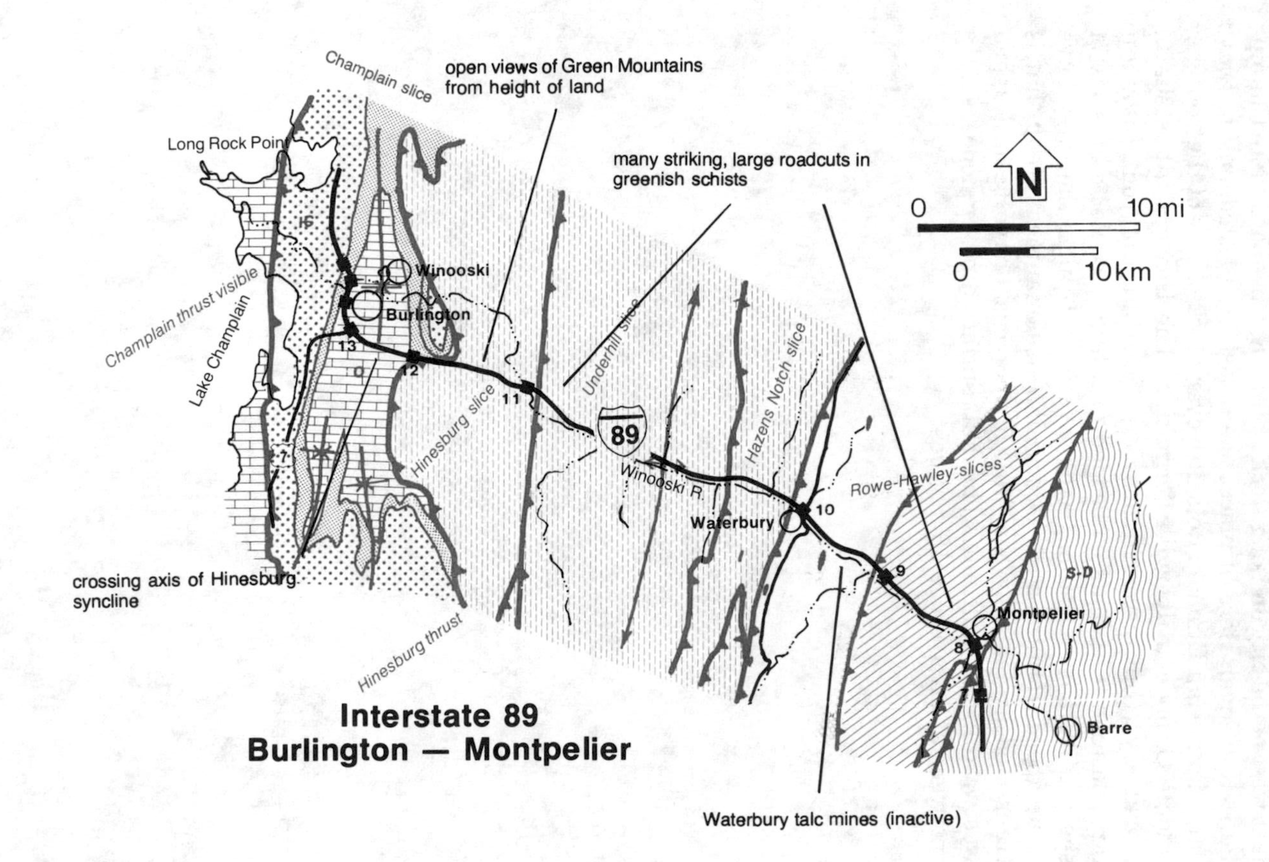

Interstate 89
Burlington — Montpelier

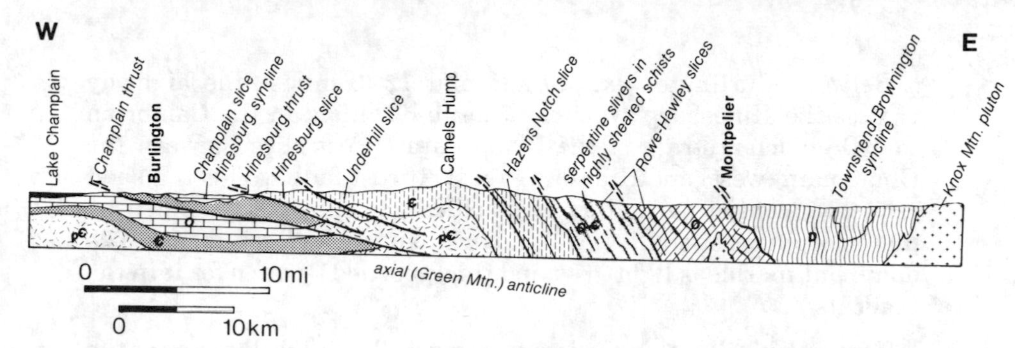

Schematic cross section parallel to Winooski River, showing major elements discussed in the text. —Adapted from C. G. Doll and R. Stanley

Interstate 89: Burlington—Montpelier

35 mi. / 56 km.

Interstate 89 between Burlington and Montpelier is geologically one of the most exciting routes in Vermont. For geologists the best thing about interstate highways is that they provide many fresh bedrock cuts that reveal the working "innards" of the earth. They give us a look below the surface we can't get otherwise, except in rare natural outcrops. Roadcuts are most useful where the highway trends across the layers, folds and faults of the bedrock, as in this case. This section traverses roughly east to west across the western thrust belt of Vermont, which trends from north to south, slicing through practically everything in the way in true interstate style. For the highway construction crews, the task was eased considerably by the preceding work of the Winooski River over millions of years in excavating its great water gap through the mountains. The roadcuts are large, numerous, and spectacular, as much for the color and pattern of the exposed rock as for the geologic history they reveal.

The landscapes of this beautiful valley illuminate an equally impressive glacial history. Most of the large roadcuts have glacially smoothed and striated upper surfaces. Much of the valley was broadly gouged by the Winooski ice lobe. In the early stages of ice recession, the Winooski Valley and several of its tributaries upstream from Waterbury were drowned in a high-level glacial lake, Lake Winooski. Later, Lake Vermont extended a long slender arm from the Champlain basin to Barre, a straight-line distance of 35 miles from its shoreline against the western flank of the Green Mountains. The result is a valley steeped in glacial drift and lake sediments, and extensively terraced through postglacial downcutting of the river.

Between Burlington exit 13 and exit 12 (3 miles), the highway crosses the Hinesburg syncline, a large downfold in the Cambrian and Ordovician strata of the Champlain thrust slice between the Champlain (west) and Hinesburg (east) thrust faults. This is a shelf sequence, consisting mainly of marbles and quartzites formed by metamorphism of shallow water, continental shelf sediments. The dominant marble is light gray and thick-bedded—watch for it in the roadcuts.

Topographically, this section is rather hilly, but still a part of the Champlain lowlands once submerged in Lake Vermont. The hilliness largely reflects deformation of the underlying bedrock.

Exit 12 lies almost astride the Hinesburg thrust fault, a profound structural boundary that places Cambrian Camels Hump schists over the marbles. The schists are dark brown in roadcuts east of exit 12, and locally striped with thin white quartz lenses. They are mostly metamorphosed graywacke, a kind of muddy sandstone from a deeper water environment that was east of the shelf sequence. Between exits 12 and 11 (6 miles), you pass several large cuts in these schists; the crumpled appearance of some hints at the intensity of deformation within the fault slice.

You cross a thrust fault of even greater significance at exit 11, this one separating the Hinesburg slice from the overlying Underhill slice. This appears to be the root zone for the Taconic slices that cap the Taconic Mountains of southwestern Vermont and constitute the Taconic klippe. There is no klippe west of here, apparently because it has been completely eroded away. This fault, and others within a narrow north to south zone, moved the Taconic slices up and over the Precambrian core of the Green Mountains, tens of miles westward. The leading edge of the slice was later "klipped" off by erosion and left stranded. Dark slates and phyllites of the Taconic Mountains are presumably the rough equivalents of the Camels Hump group rocks in the Underhill slice, but they were metamorphosed at lower temperatures.

The height of land between exits 12 and 11 provides splendid views of the Green Mountains. People traveling east first enter the Winooski Valley near exit 11, where flat river terraces are visible upstream. The modern floodplain, which is here very broad, will itself be dissected as the river continues to erode downward, and parts of it may also be preserved as terraces in the future. If you search carefully, you will see many more terraces in the section of valley between exits 11 and 8 to Montpelier (26 miles). Sand and gravel pits on the valley sides commmonly reveal terrace deposits that are not otherwise obvious, and some are at surprising elevations above the modern

river; one such is near mile marker 73, 6 miles east of exit 11. It is hard to believe that Lakes Winooski or Vermont were so deep.

Between exit 11 and South Duxbury, you cross from the base to the top of the Underhill slice, still in Camels Hump schists. Most are attractive greenish biotite schists in which the color comes from the soft, micaceous mineral chlorite. Many also contain metamorphic garnet, and nearly all reveal abundant white quartz lenses between the layers.

At South Duxbury, the road crosses yet another major thrust fault. In the next 5 miles to exit 10, roadcuts expose brownish, more biotite-rich, schists of the Hazens Notch thrust fault slice. The slice in northern Vermont is more strongly sheared and contains many serpentine slivers, and thus is part of the serpentine belt.

Between exits 10 and 8 (11 miles), you cross the main serpentine belt of the Rowe-Hawley slices. Most of the rocks are thinly-leaved phyllites with lustrous cleavage surfaces. They began as sediments and volcanic debris scraped from the deep ocean floor on the continent side of the Bronson Hill island arc, in the early stages of Taconian mountain-building, then piled up westward against the continental margin. The serpentine slivers are pieces of dark crust ripped from the ocean floor in the process and moved upward along thrust faults. The original dunites, peridotites, or gabbros were partially or totally altered to serpentine and related minerals during metamorphism. None of the serpentine pods are visible in Interstate 89 roadcuts. However US 2 passes close by the Waterbury talc mine southeast of Waterbury, where you can find serpentine and related minerals in the mine dumps.

Interstate 89
Montpelier—White River Junction

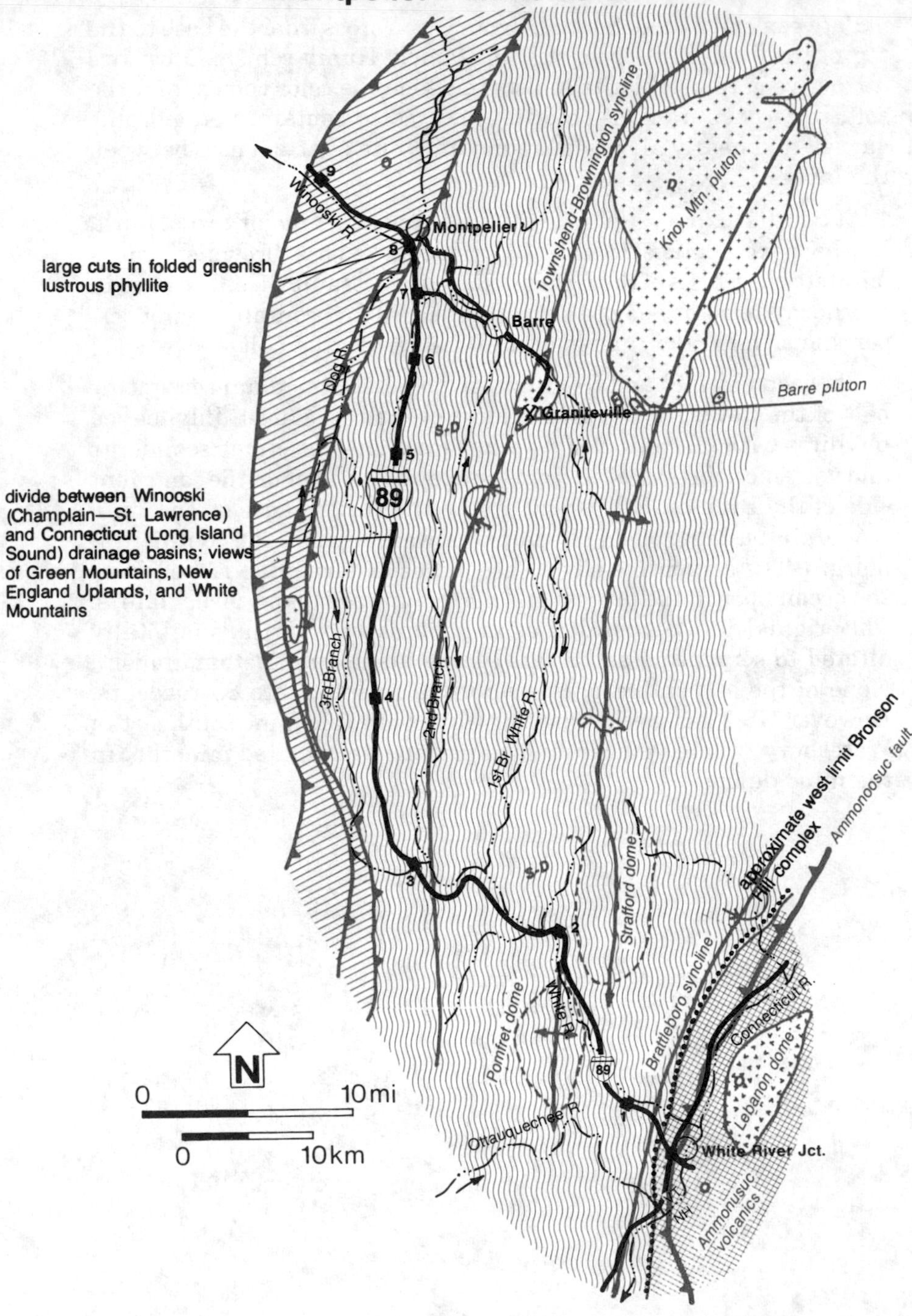

Folded mica schist of Missisquoi formation near Montpelier.

Interstate 89: Montpelier—White River Junction

52 mi. / 83 km.

Immediately south of exit 8 to Montpelier, the road climbs out of the Winooski Valley past enormous roadcuts in the Barton River member of the early Devonian Waits River formation. Exposed are phyllites with lustrous cleavage surfaces and interbedded marbles. A large downfold is well defined by the layering on the northeast side of the road. Most of the roadcuts to White River Junction expose this formation.

The divide between the Champlain drainage basin, the Winooski River, and the Connecticut River basin, the White River, lies at about exit 5 (10 miles). Between there and exit 3 (23 miles), the road continues almost straight north-south along the crest of the ridge that separates branches of White River, with many open views of the Green Mountains to the west and the Piedmont, or New England uplands, section of Vermont to the east. The entire traverse from Montpelier to Bethel nearly parallels the steeply dipping strata of the

Lustrous cleavage surfaces of phyllitic schist near Montpelier.

east limb of the Green Mountain anticline, the west limb of the adjacent Townshend-Brownington syncline.

Three miles north of Bethel, near exit 3, is the site of the attractive, white Bethel granite. The rock locally contains peculiar dark knots called orbicules, rich in biotite mica and shaped like prunes. Several years ago in a study of identical rocks from Craftsbury, Vermont, I found that the mica flakes are arranged in concentric shells like wet leaves around a tennis ball with quartz, feldspar, and other minerals dispersed among them. At Craftsbury, the orbicules are crowded near what appears to be the ceiling of the magma chamber, where the original molten granite was in direct contact with the solid country rock of its container. The orbicules there were flattened, as you might expect a soft tennis ball covered with wet leaves to flatten if it were pressed against the roof of a magma chamber. These features suggest that the orbs formed on gas bubbles that rose through the melt in which early-formed biotite crystals were suspended. The biotite flakes the bubble met along the way simply clung to it by surface tension. The longer the upward journey, the thicker the accumulation of shells and entrapped liquid. All of the Craftsbury orbicules were more or less flattened.

Interstate 89 crosses White River once near exit 3 and two more times between exits 3 and 1 (21 miles). The route traverses diagonally southeast across the Townshend-Brownington syncline, Pomfret dome, and Brattleboro syncline. Rocks near the Pomfret dome are more gneissic and of higher metamorphic grade, recrystallized at a higher temperature than most of the rocks traversed so far. This may result from contact metamorphism associated with intrusion of an igneous body at shallow depth under the dome.

Near White River Junction, the Connecticut Valley is open to view. Location of the river here appears to be controlled by the Ammonoosuc fault that can be traced for about 80 miles farther north and 3-4 miles south of White River Junction. Glaciation has greatly modified the Connecticut Valley, first by erosion during southward advance of ice, later by sedimentation and erosion during ice retreat. It is the site of glacial Lake Hitchcock, which at its maximum, reached from Middletown, Connecticut, to Canada, spanning the Vermont-New Hampshire border from one end to the other. The White River Valley was one of the main branches of the lake, which extended to a point about midway between exits 4 and 5. White River Junction lies almost on top of the western margin of the Bronson Hill island arc complex, perhaps the single most important element in the plate tectonics interpretation of Vermont and New Hampshire. Collapse of the volcanic islands against the continent during Taconian mountain-building supplied the compressive force for all of the great thrust faults of western Vermont traversed in the Burlington-Montpelier segment of Interstate 89. The buckling of the Devonian strata, just traversed, into anticlines and synclines obviously occurred later because the rocks are younger than Taconian mountain-building. They were folded during the Acadian event when the proto-Atlantic closed and ancestral North America collided with Europe.

The enormous cuts on the Interstate 89-Interstate 91 interchange at White River Junction are greenschists of the Post Pond volcanic member of the Ordovician Orfordville formation, presumably a part of the collapsed island arc.

Biotite orbicules in granite from Craftsbury, Vermont, similar to those of the Bethel granite.

terracing, abandoned meander loop, and waterfall on the Mascoma River

Ammonoosuc fault

Bronson Hill complex

White River Jct.

Lebanon

Mascoma dome

Oliverian

Ammonoosuc volcanics

Mt. Clough pluton

view west to bare granite summit of Croydon Dome

Grantham

Croydon dome

Cardigan pluton

3 foot black basalt dike

strikingly porphyritic Kinsman quartz monzonite; many xenoliths of mica schist

Merrimack R.

old quarries in Concord granite

Hopkinton

Concord

0 10 mi

0 10 km

N

Concord pluton

Interstate 89
White River Junction — Concord

Interstate 89: White River Junction, Vt.— Concord, N. H.

59 mi. / 94 km.

In the five miles between White River Junction, Vermont, and Lebanon, New Hampshire, Interstate 89 crosses the southern tip of the Lebanon dome, an anticline shaped like an overturned bowl. This is just one of numerous similar structures arranged like a string of sausages along the axis of the Bronson Hill anticline in western New Hampshire. Within the state, the elongate anticline stretches for 150 miles from the Massachusetts border to the Maine border northeast of Berlin, and averages about 15 miles wide. It continues southward across Massachusetts and Connecticut to Long Island Sound. The structure appears to be a collapsed and metamorphosed volcanic island arc formed in early Taconian time as the proto-Atlantic Ocean basin began to close. Each of the many domes is a bulge in the larger anticlinal axis. Each is cored by an oval mass of gneiss and granitic rocks that belong to the so-called Oliverian plutonic series of apparent Ordovician age.

Crustal convergence at the beginning of Taconian mountain-building ruptured the basaltic ocean crust east of the ancestral North American continental shelf and shoved the western slab under the eastern. When the sinking slab reached deep into the mantle, heat drove water out of the rocks to form red hot steam that rose into the already hot rocks above. They partially melted to form the magmas that became the volcanic rocks and the igneous intrusions within them. Most of the domes are enveloped by the Ammonoosuc volcanic rocks that they intrude. The Lebanon dome, instead, is enclosed in Orfordville schists derived from original sedimentary rocks that are older than the Ammonoosuc formation.

The Lebanon dome also differs from the other domes in that it is not along the Bronson Hill anticlinal axis, but some 8 miles west of it, between the west-dipping Ammonoosuc and Northey Hill thrust faults. It is considered a member of the Oliverian series mainly because the rocks and general geologic environment are similar. The rocks of the series are distinguished in the field from the late Devonian New Hampshire plutonic series by their more extreme granulation and pink color. They are also more granulated than the rocks of the White Mountain plutons.

Roadcuts near Lebanon expose a dark gray gneiss of a narrow border zone that surrounds the granitic core of the dome. In places the gneiss contains whitish fingers and lenses, or veins, of granite.

The Mascoma River enters the Connecticut River just south of the White River juncture. Like most of the tributaries of the Connecticut River, the Mascoma River is entrenched, a result of postglacial downcutting. East of Lebanon, an abandoned meander loop and waterfall mark a former position of the river about 60 feet above its present level.

Between exit 18 to Lebanon and exit 16 (5 miles) are several roadcuts in Orfordville formation schists. Most are dark-gray slates derived from original clay muds, or greenschists derived from original basalts. Several cuts have very rusty surfaces from oxidation of iron pyrite contained in the rock. A few whitish quartzites and metamorphosed conglomerates formed from sands and gravels. Nearly all of the layering is steeply inclined to vertical. Closer to the Lebanon dome, the Orfordville formation has been altered to a more coarse-grained grayish gneiss, possibly as a result of contact metamorphism along with introduction of chemicals from the intruding granite magma, a process called metasomatism.

Between exits 16 and 13 to Grantham (9 miles) are several very large roadcuts in highly deformed rocks, including 3-4 units of different age and composition. They include dark Devonian schists, Silurian marbles and quartzites, Ordovician volcanic rocks, and some dark gray Bethlehem gneiss that belongs to the Mount Clough pluton of the New Hampshire plutonic series. The complexity of deformation in this section appears to result in part from intrusion of the Mount Clough pluton at the same time as the most severe folding, shearing, faulting, and metamorphism that accompanied Acadian mountain-building.

The rather bare summit of the Croydon dome, another Oliverian granite pluton, is visible west of exit 13 to Grantham.

You cross the Mount Clough pluton between exits 13 and 12 (8 miles). Several large roadcuts expose massive gray homogeneous

Black basalt dike in jointed granitic gneiss at exit 12, 32 miles northwest of Concord.

gneiss, in places shot through with thin whitish granite veins, elsewhere cross-cut by thick white pegmatite dikes that contain large sheets of white muscovite mica. At exit 12, a 3-foot wide black basalt dike intruded along a joint in a section of highly jointed gneiss.

Between exits 12 and 9 (16 miles) you cross the Cardigan pluton of the Kinsman quartz monzonite, a variety of granite. It is one of the most striking rocks of New Hampshire, also part of the New Hampshire plutonic series. Much of the rock is porphyritic, full of oversized rectangular crystals of white feldspar up to 3 inches long, that float in a finer-grained gray matrix rich in biotite mica. In places the crystals are well aligned. Locally, the rock contains large red garnets up to an inch or more in diameter; in some places these are partially or totally replaced by biotite mica, but the six-, or so, sided outlines of the original garnets are preserved. Exotic metamorphic minerals such as cordierite and sillimanite are also quite common, but may be detectable only under the microscope.

Slab-like inclusions of mica schist are everywhere in the Kinsman quartz monzonite. They range in length to thousands of feet. Many are ghost-like, having been engulfed in magma and altered by reaction with it to a rock resembling the igneous host. In places, inclusions are so numerous and entangled with the host rock that the mixture has been mapped as a separate and distinct unit. As in the Mount Clough pluton, these features strongly suggest that the Kinsman quartz monzonite was also intruded during the height of Acadian mountain-building. Severe deformation accounts for the extensive fracturing of the intruded schists and engulfment of separate blocks by the invading magma. In places, the inclusions are disoriented and deformed. Elsewhere, the granitic rock is itself extensively fractured, or brecciated, the fragments suspended in an obviously later injection of magma.

Between exits 9 and 3 (19 miles), you cross a broad swath of mica schist and micaceous quartzite that belongs to the early Devonian Littleton formation. These rocks also contain the mineral sillimanite, indicating very high-grade metamorphism at high temperature. The route from White River Junction to this point crossed metamorphic zones from chlorite, to biotite to garnet to staurolite to sillimanite. These minerals always crystallize in a certain temperature-pressure environment of metamorphism in rocks of appropriate composition, mainly shales formed from original clay muds. Their succession indicates increasing temperature—progressively higher grade metamorphism from west to east.

Between exits 3 and 1 to Concord (5 miles) are several roadcuts in the Concord gneiss, another member of the New Hampshire plutonic series, but younger than the Kinsman quartz monzonite. In places, where the two units are together, veins and dikes of Concord gneiss

Kinsman quartz monzonite showing large rectangular crystals of potash feldspar that weather in relief, 24 miles northwest of Concord.

intrude the Kinsman quartz monzonite. The Concord pluton is an oval body, about 5 miles long from north to south and 4 miles from east to west, that intrudes the Littleton schists. Several other, much smaller, plutons exposed in a 10-mile radius around Concord are probably offshoots of the same granitic source. Many superb exposures of this unit are in quarries around the state capitol. The rock has a remarkably uniform composition and texture that makes excellent dimension stone for building, tombstone, or sculpture stone, for which New Hampshire is world-famous. Here the rock is generally light grayish, fine-grained and homogeneous.

Concord quarries have supplied granite for many famous buildings, including the Library of Congress in Washington, D. C. New Hampshire granite was also used as the cornerstone of the main United Nations building in New York City.

Interstate 91: The Connecticut River Valley

The Connecticut Valley marks an important plate tectonic boundary as well as the political boundary between Vermont and New Hampshire. It also contains a wealth of depositional and erosional features related to Wisconsin glacial recession. The Connecticut Valley is historically important as a corridor for settlement of New England, and it is one of the most beautiful regions of the northeast.

Between the Massachusetts border and Lancaster, New Hampshire, the river follows the western margin of the collapsed Bronson Hill island arc complex, that formed in the early stages of the Taconian mountain-building cycle and later collided with the continent. The river thus separates two radically different terrains, the ancient basement and younger metasedimentary cover of ancestral North America on the west and island arc on the east. If that interpretation is correct, much or all of New Hampshire east of the narrow arc complex is "exotic" terrain, formerly a piece of Africa or Europe.

South of the border in Massachusetts the river occupies a somewhat different bedrock environment where it meanders over the floor of the Connecticut Valley basin. This is one of many linear fault-block basins near the east coast of North America that contain distinctive red sediments and basalt lava flows of Triassic-Jurassic age. All of the basins are products of crustal stretching that began about 200 million years ago and eventually led to the separation of the Americas from Eurasia and Africa, with the opening of the Atlantic Ocean basin. This happened more than 200 million years after the collapse of the island arc that became the Bronson Hill complex.

Most of what we know today as the Connecticut Valley was here before Wisconsin glaciation. Ice crowded into the valley and moved southward along it, gouging it deeper and wider. As the glacier finally melted, the valley held a lake, called Lake Hitchcock, that

Varved clay at Putney. Keys for scale. Note the boundaries between varves with a thin dark layer at top of each.

gradually lengthened upvalley as the ice receded northward and at one point stretched from Middletown, Connecticut, to the Canadian border. Kame terraces formed as "strip deltas" alongside the ice as it retreated.

Copious quantities of sediments released from the melting glacier and washed from the barren lands newly freed of ice piled up under water on the valley floor. The deposits include the famous varved clays that have yielded so much information on the timing of glacial recession from this region. Each lamination in the varved clays grades from light to dark gray upward and represents one year's accumulation. The lighter material is rock flour released from the melting ice during the summer; the darker, thinner layers built up under the lake ice in winter.

Studies of the Connecticut Valley varves by Ernst Antevs (1922) revealed that it took 4300 years for the Wisconsin ice front to recede from Middletown, Connecticut, to St. Johnsbury, Vermont, a distance of about 200 miles and an average recessional rate of about 245 feet per year. Much glacial study in New England has focused on varved clay because there are so few moraines to mark various stands of the ice terminus.

Working out the timing by Antevs's method assumes first that

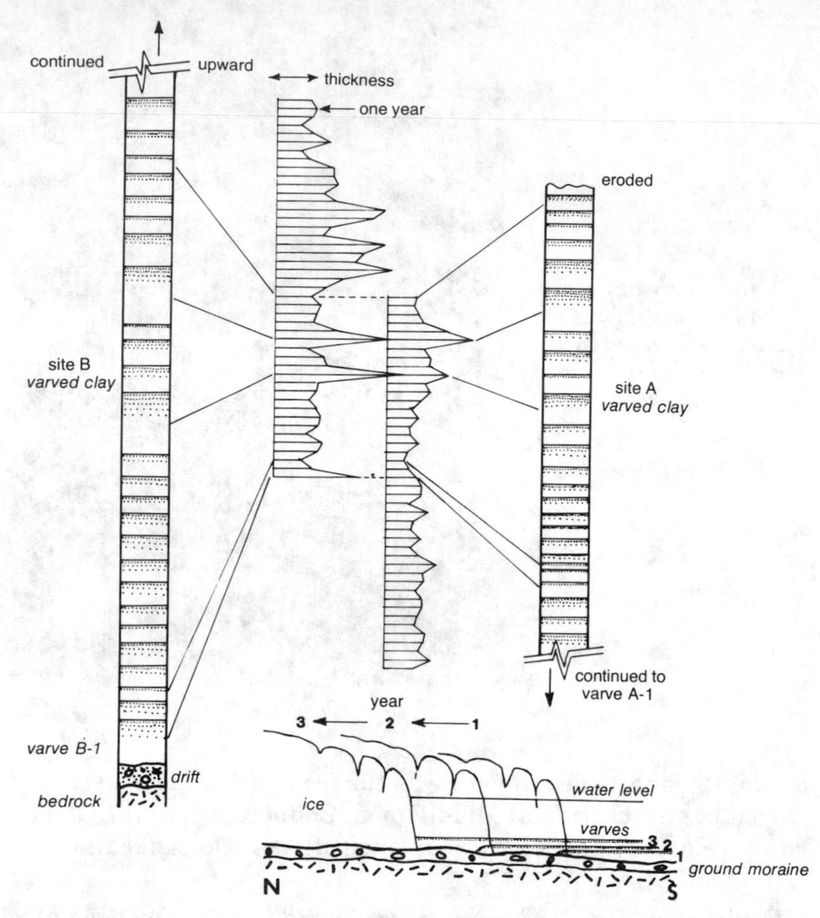

Correlation of varves from one location to another is accomplished by measuring year-to-year variations in thickness. Varves everywhere are thicker in warm years when more sediment is released in meltwater. The first year varve at a site is deposited only after that site is free of ice as illustrated in inset.

varves were constantly being formed as the ice front moved northward. Thus, the earliest varves formed in the south, while the land to the north was still under ice. Secondly, it assumes that varves formed in the same sequence of years can be matched from one exposure to another. Only in this way would it be possible to total up the years involved because the varved records are everywhere only fragmental. The method also assumes that the varves would everywhere be thicker in warm years when more rock flour sediment would be shed from the melting ice. Individual varves, thick or thin, could not, by themselves, be matched from place to place; but records showing thickness fluctuations over scores of years could be. So Antevs simply made graphical records of each exposure, showing varve thicknesses measured one by one from the base to the top of each exposed section. Then he placed pairs of graphs side by side, and shifted them until he found matching segments. When records overlapped or correlated only partially, the earlier years of one and later years of the other could be added on to the overlapping part. The lowest varve at a particular site marks the first year of accumulation after the site was freed of ice. By repeated correlations, Antevs was able to construct a continuous record.

Most rivers in the glaciated regions of North America have eroded their channels deeper during postglacial time. The enormous mass of the glacier deeply depressed the land, but it sprang back up after the ice melted, thus causing the rivers to cut downward. The amount of rebound increased northward where the ice was thickest, and it was minimal at the southernmost stand of the glacier margin, where the ice was thinnest. The plain of Lake Hitchcock has been particularly useful in charting the rebound in Vermont and New Hampshire. The method involves measuring the present altitudes of numerous preserved Hitchcock shoreline features such as beaches, sand bars, and wavecut benches, all of which formed at the same level. When these

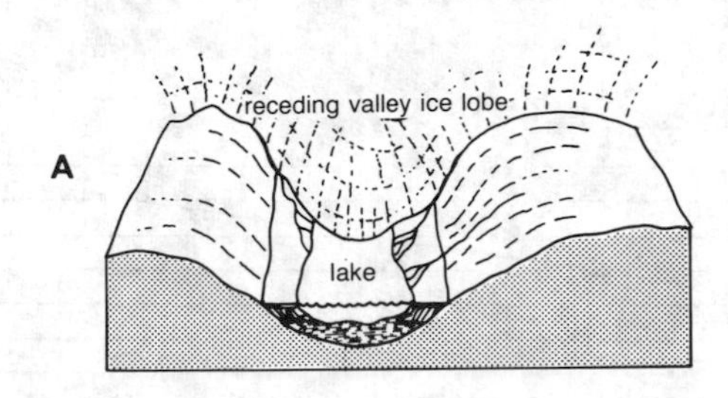

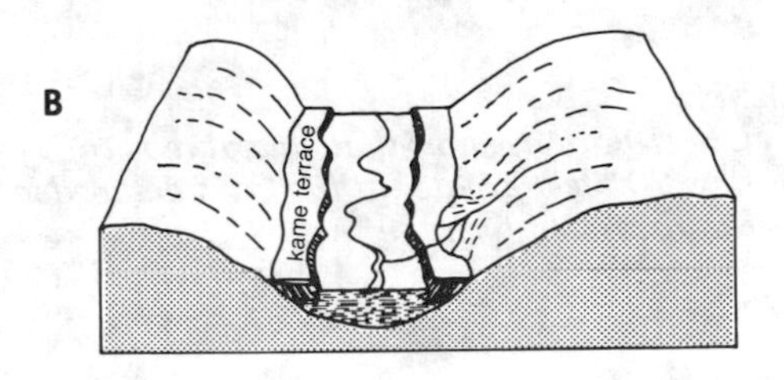

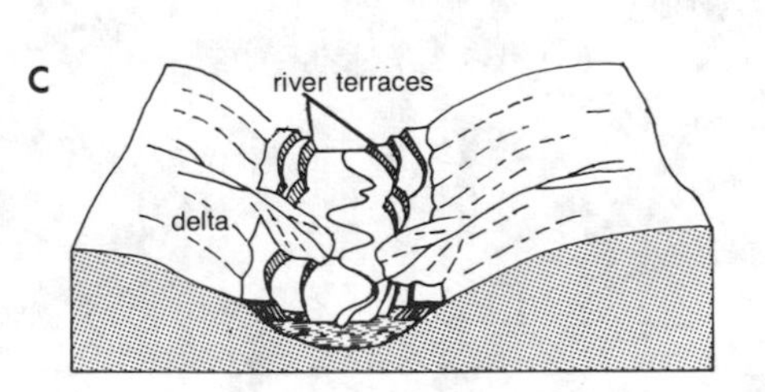

Sequence of drawings showing formation of kame terrace by ice marginal stream and filling of valley with proglacial lake sediments. Postglacial uplift or reduction in sediment load, causes downcutting of river and carving of river terraces.

were graphed according to elevation with the proper horizontal spacing between them, a profile of a tilted lake plain emerged, in which the northern end at the Canadian border is now 700 feet higher than the southern end at the Massachusetts border.

The river cut into the valley fill during the rebound, creating river terraces as it meandered back and forth, and exposing cross sections of varved clays and other valley-fill sediments. The flat-topped terraces are now visible in many parts of the valley, although vegetation makes many of them difficult to identify.

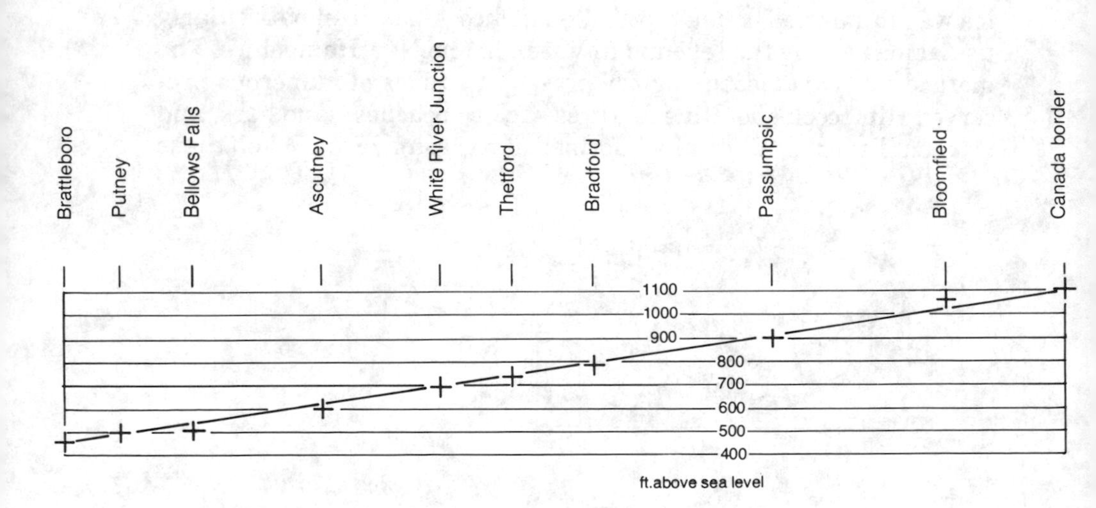

Profile plot of Lake Hitchcock shoreline elevations show that glacial rebound of the originally level lake plain is nearly 700 feet greater at the Canada border than at Brattleboro. From Stewart and McClintock, 1969.

Interstate 91: Massachusetts Border— White River Junction, Vt.

70 mi. / 112 km.

Between the Massachusetts border and Brattleboro (8 miles), several roadcuts expose the Devonian Littleton formation. The rocks are generally rather dark gray phyllites with lustrous cleavage surfaces formed by low-grade metamorphism of shales with thin interlayers of white quartzite formed from sandstone. The Connecticut River in this section is a few miles to the east and out of view.

North of Brattleboro, the highway remains close enough to the river for more than 100 miles to provide many good views of the valley. The valley floor at river level is locally quite broad and everywhere very flat. This is the flood plain of the modern river on which it continually meanders back and forth. It is also working downward into the valley fill, although the rate of downcutting is very slow. As the river works its way deeper, the active floodplain will be carved at an ever lower level, but some of the current floodplain may survive the erosion as a terrace. This has been going on for thousands of years, since some time after Lake Hitchcock was drained; and that's why there are now river terraces at several levels on the valley sides in addition to the surviving kame terracing above them. The road lies atop the terraces in many places; and many villages, including Brattleboro, are built on them, or on the floodplain. It will take a little practice to recognize the terraces. They are most conspicuous where they have been cleared for cultivation. The higher ones, being older, are often more deeply dissected by tributary streams than the lower ones.

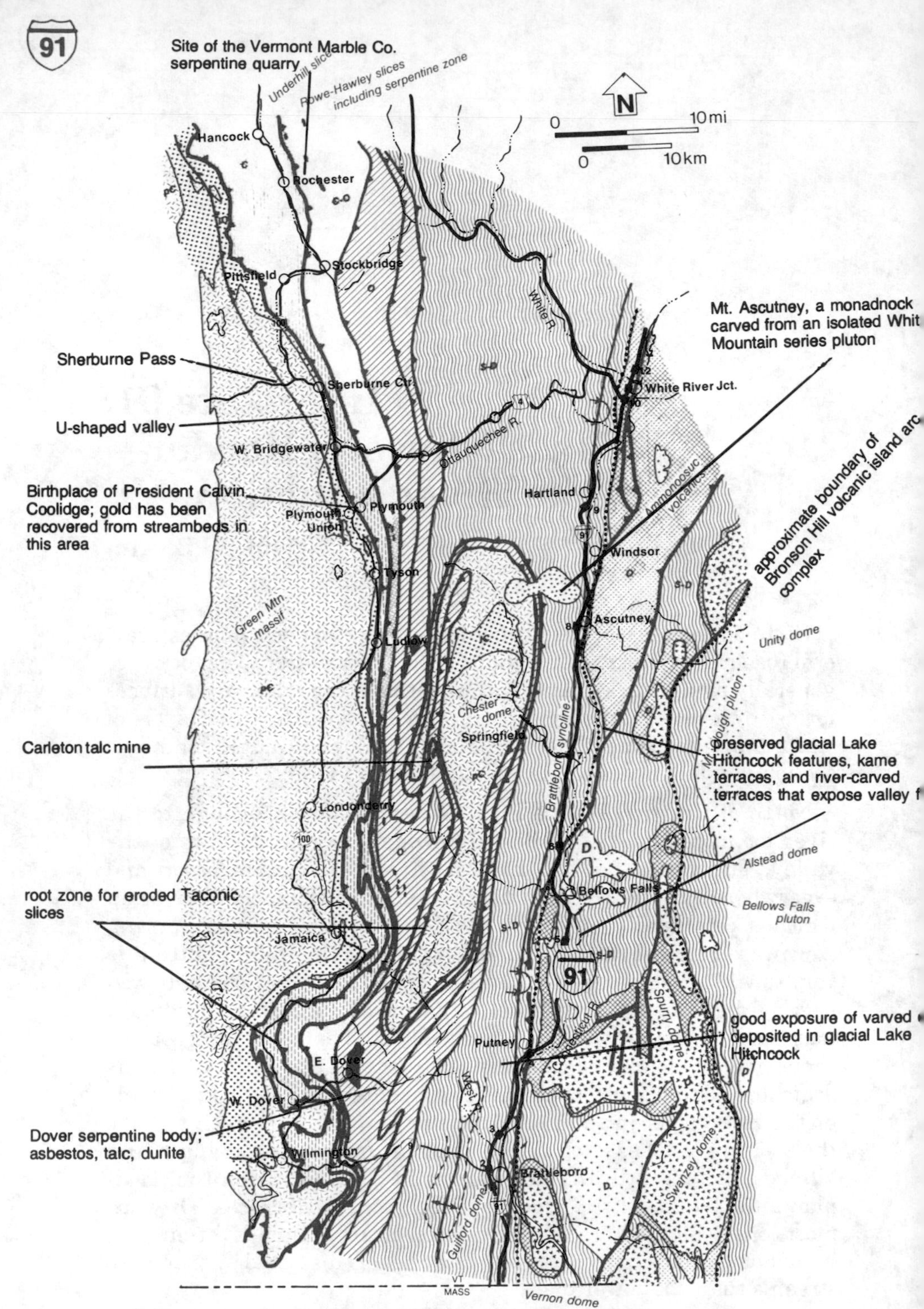

Interstate 91, VT 100
Massachusetts Border — White River Junction

As you drive this route, note that most of the small tributary streams are entrenched into narrow, steep-sided valleys. This is also a product of the post-glacial downcutting of the main stream. In general, the Connecticut River has controlled the leveling of the land because it carries the most water; and is thus the base level for regional erosion. The tributaries keep the downward pace, but have so little energy to cut sideways that the results are narrow valleys. A good example is Mill Brook by exit 4 to Putney; but there are many more. The entrenched character is not always obvious close to the master river; you need to look up the valley where the streams have cut deeply into bedrock, in some places carving narrow gorges. Shorelines of glacial Lake Hitchcock extend far up into West River valley which the highway crosses by exit 3 north of Brattleboro.

A good exposure of varved clay at Putney, is in the banks of the small creek that intersects US 5 by the entrance to Interstate 91. This clay so far from and so high above the modern river gives a good impression of the large size of Lake Hitchcock.

At Putney (exit 4), you cross the axis of an anticlinal fold with roadcuts exposing the Ordovician slates and phyllites of its core. The rocks resemble those of the Littleton formation, but lack the white quartzite interlayers. Between Putney and Bellows Falls (exit 5—10 miles) are cuts in gently dipping, or sloping, layers of slate and phyllite near the crest of a large anticlinal fold that trends from north to south. Elsewhere the layers of the fold limbs are steeply inclined. Actually, this fold is only a wrinkle on the much larger Brattleboro syncline. The syncline consists of tightly folded strata between the collapsed Bronson Hill island arc complex, and the domes of eastern Vermont, upfolds shaped like overturned bowls. This, in turn, is only a small part of the much larger Connecticut Valley-Gaspe syncline that extends into Canada. Simply put, the rocks of the Brattleboro syncline were caught in the vice between the great masses of the ancestral continent and the island arc as the proto-Atlantic basin closed. The constriction is narrowest between Putney and Ascutney (exit 8—34 miles), where the Athens dome on the south, and the Chester dome on the north, each cored with Precambrian basement rocks, lie only 5-10 miles west of the river. Note the roadcuts in this section display somewhat more severe deformation.

Mount Ascutney, at 3144 feet with the tower on top, is the prominent summit visible from many points between Bellows Falls (exit 4) and Pompanoosuc (8 miles north of White River Junction exit 12), a distance of 50 miles. The highway skirts the base of the mountain just north of Ascutney (exit 8). A paved road in Mount Ascutney State Park leads to the summit and provides scenic views of the surrounding country, including the Connecticut Valley. The mountain is an isolated peak that stands nearly 2000 feet above the surrounding terrain, and is held up by a small, isolated igneous pluton of the White Mountain magma series. Practically all of the other members of this series are in the White Mountains of New Hampshire. Here the rocks are of variable composition, ranging from dark gabbros to pale granites. The relatively small Cuttingsville pluton, which intrudes the Precambrian core of the Green Mountains 20 miles west of Mount Ascutney, is another small member of the White Mountain series exposed in Vermont.

The village of Windsor, on a river terrace 5 miles north of Ascutney, is called the birthplace of Vermont. The state constitution was signed there on July 8, 1777. Meetings of the General Assembly were held there until Montpelier was established as the permanent capital in 1805.

Between Ascutney and White River Junction (18 miles) are many large roadcuts in phyllites and mica schists of the Devonian Gile Mountain formation, most of them with nearly vertical lustrous cleavage. The North Hartland Dam on the Ottauquechee River is visible west of the highway near White River Junction. This section of the Connecticut Valley is particularly broad, well-terraced, and dotted with many, very large sand and gravel quarries. Erosion of one 2-mile long segment of terrace near Hartland left a ridge of sand and gravel that rises 60-80 feet above the highway and separates it from the river.

Highly sheared and folded schist on Interstate 91 at mile 77. Note the lenses of white feldspar–rich rock that have been stretched and broken as a result of the shearing.

Interstate 91: White River Junction—Barnet

50 mi. / 80 km.

The small size and modest character of the village of White River Junction belie its importance as a major juncture where roads, railroads, and rivers converge. It is an eloquent reminder of the close relationship between human activity and geology, particularly in the early settlements. River courses everywhere are largely controlled by the underlying rocks and, to the early settlers, the rivers often offered the easiest travel routes. Before railroads, the Connecticut River was the main artery for commerce in Vermont and New Hampshire; and White River Junction, situated by its largest tributary, naturally became one of its most active villages. Paradoxically, although a pleasant stopping place, it remains little more than a juncture today; it has not grown into a big city; and most travelers only pass through on their way to other places.

The White River valley and its tributaries held the longest branch of Lake Hitchcock during deglaciation. Lakeshore features marking the same water plain you can see here are present more than 40 miles upstream from the Connecticut River, reaching to East Brookfield in its northernmost extension along the second branch of the White River. This is only 5 miles from the headwaters of Stevens Creek, tributary to the west-flowing Winooski which held an arm of Lake Vermont that reached inland from the Champlain basin. The two lake basins are separated by only a low drainage divide. If Lake Hitchcock had been just a little bit higher, it might have reached most of the way across the state.

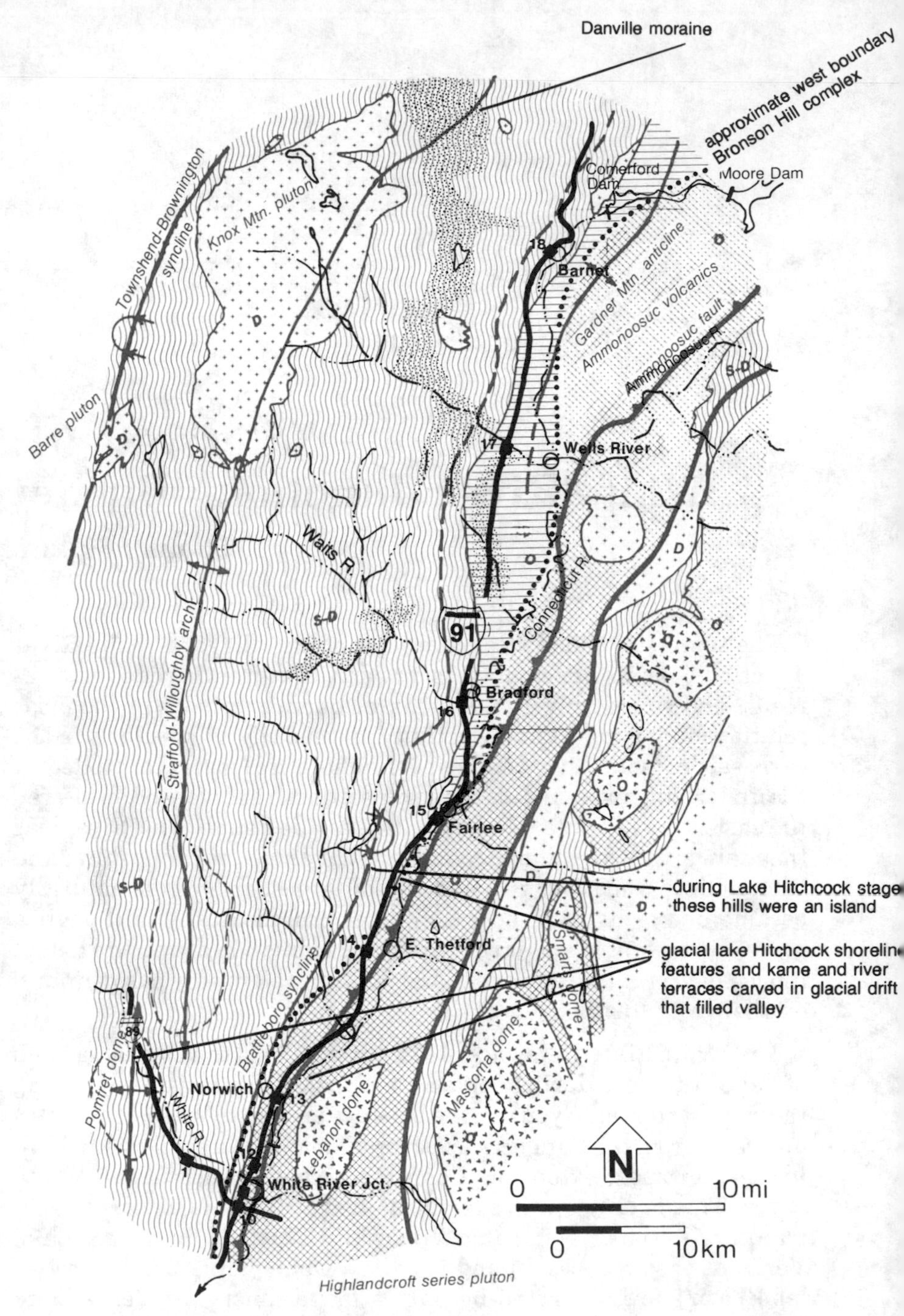

Interstate 91
White River Junction — Barnet

Roadcuts near the White River crossing expose granite gneiss in a small body of the Ordovician Highlandcroft plutonic series intruded into an anticlinal fold. The intruded rocks are mostly Ordovician Post Pond formation greenschists, apparently derived from original volcanic rocks of the island arc complex. The same dark greenish rocks are exposed in several cuts between White River Junction and Thetford (14 miles); the deeply weathered surfaces that survived blasting are rusty because of the oxidation of iron-bearing minerals. This section is also alongside the Ammonoosuc fault, and the intense shearing and folding visible here may, in part, result from displacement along the fault. Post Pond volcanic rocks on either side of the fault contain different metamorphic minerals. Those east of the fault contain garnet that indicates a higher metamorphic grade than in the rocks immediately west of the fault, in which the mineral chlorite grew during low-grade metamorphism. The two rock masses were metamorphosed in quite different and probably widely separated environments, and then movement along the fault brought them together.

Hills immediately west of East Thetford were part of an island in Lake Hitchcock. A branch of the lake extended up the Ompompanoosuc Valley, which joins the Connecticut Valley south of here at Pompanoosuc. The lake wrapped around the hills and rejoined the main body of water 4.5 miles north of here by the village of Ely. The lower Ompompanoosuc River is now impounded behind Union Village Dam, and Lake Fairlee occupies the northernmost segment of the loop. The Lake Hitchcock shoreline has been traced throughout the loop.

Between East Thetford (exit 14) and Fairlee (exit 15—7 miles), Interstate 91 lies close to the Connecticut River where the floodplain is particularly broad, the river meanders widely, and the terraces are well-defined. Just north of exit 15 is an enormous cut and cliffs of "The Palisades" of Morey Hill constrict the valley. The hill is supported by another body of light gray granite similar to the one at White River Junction. The Ammonoosuc fault traverses the base of the cliff, and the topographic relief suggests that the west side, with the granite, moved up. The southern view of the valley from mile 95 just north of the big cut is quite scenic, owing in part to the presence of the steep ramparts on either side of the constriction.

Between Fairlee and Wells River (exit 17—19 miles), the road lies mostly away from the river, except near Bradford (exit 16), at the Waits River juncture. Numerous large roadcuts in this segment are mostly in quartz-rich phyllites, greenschists, micaceous quartzites, and slates of the Ordovician Albee formation. Before they were metamorphosed, those rocks were shaley sandstones, shale, and light-colored volcanic rocks. The micaceous quartzites contain distinctive alternating dark and light "pinstripes."

The highway passes over Wells River just north of exit 17. Beginning only a short distance upstream are exposures of the most extensive moraine belt in northern New England. This Danville moraine is a nearly continuous ribbon-like deposit that stretches for about 50 miles in roughly a south to north direction from near Newberry, 6 miles south-southeast of exit 17 by the Connecticut River, almost to Barton at exit 25. The moraine marks the stationary position of the ice terminus for a long time during Wisconsin glacial recession which proceded more or less in a northwesterly direction. The till as a whole is largely free of clay and very sandy, making it ideal for road construction. There are many quarries along the length of the moraine.

Between Wells River and Barnet (exit 18—10 miles), the road is close to the river with lovely overviews of the valley; the White Mountains are visible to the east. Most of the many large roadcuts are phyllites of the Albee formation. About three miles south of exit 18, you cross the unconformity that separates the Albee from the much younger Devonian Gile Mountain formation. Cuts near Barnet expose the grayish micaceous quartzites and schists of this latter unit. The unconformity represents an erosional gap in the geologic record of perhaps 30-50 million years, with rocks formed during the Silurian period of 430-395 million years ago completely missing.

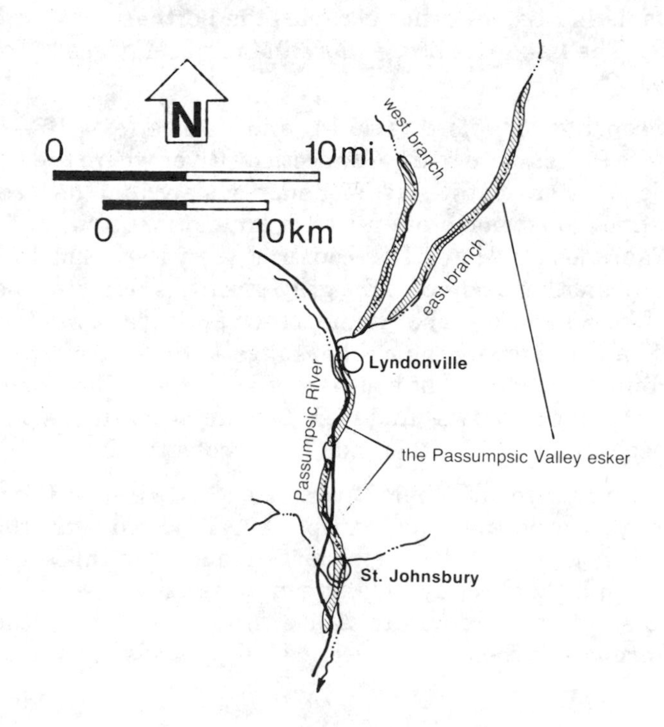

Sand and gravel quarry in the Passumpsic Valley esker north of St. Johnsbury. These clean, water-washed esker deposits are a valuable natural resource.

Interstate 91: Barnet—Derby Line

58 mi. / 93 km.

This segment of Interstate 91 angles across the Connecticut Valley-Gaspe (Brattleboro) syncline. Many roadcuts expose phyllites, schists, and micaceous quartzites of the metamorphosed Devonian Gile Mountain formation and marbles, phyllites, and schists of the Devonian Waits River formation, also metamorphosed. The northern part of the route weaves among several large bodies of granite, part of the Devonian New Hampshire plutonic series. Rocks near the granites were subjected to higher temperatures and so contain different metamorphic minerals; they have been contact metamorphosed.

Between Barnet (exit 18) and Lyndonville (exit 23—17 miles), the highway follows the Passumpsic Valley. The phyllites, exposed in roadcuts between St. Johnsbury and Lyndonville contain thin interlayers of dark green amphibolite, metamorphosed basalt. The most interesting geologic feature in this segment is the Passumpsic Valley esker. The southern tip of the esker ridge is near exit 19 to St. Johnsbury. From there, it continues upvalley to Lyndonville where it splits into two branches. The west branch, in the west branch Passumpsic Valley, extends northward to West Burke, and the east branch, in the east branch Passumpsic, to East Haven. The longer

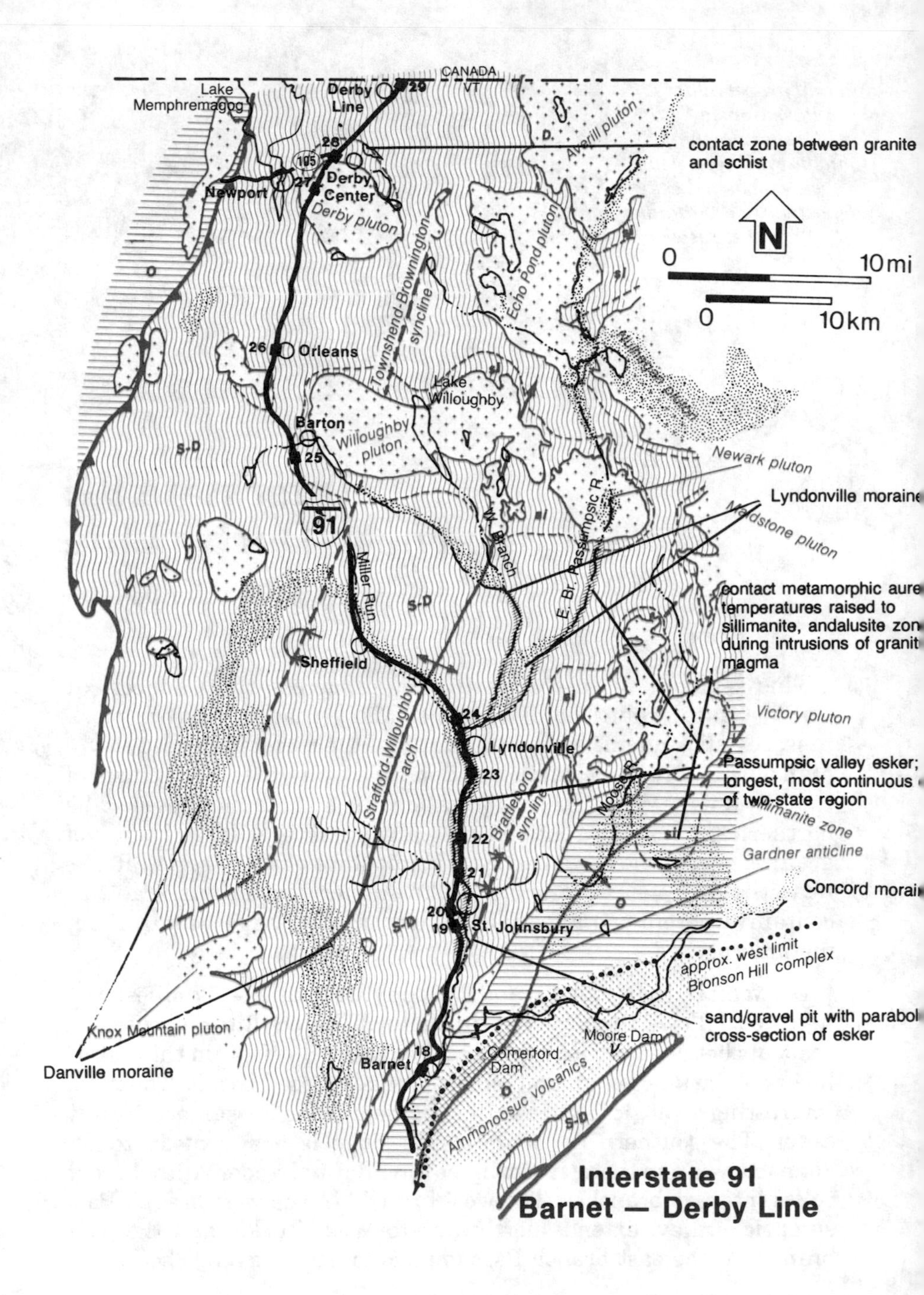
CANADA
VT
Lake Memphremagog
Derby Line
29
28
105
27
Newport
Derby Center
Derby pluton
Averill pluton
contact zone between granite and schist
N
0
10 mi
0
10 km
Echo Pond pluton
Townshend-Brownington syncline
26
Orleans
Lake Willoughby
Barton
Willoughby pluton
25
S-D
Newark pluton
Lyndonville moraine
Maidstone pluton
91
Miller Run
W. Branch
E. Br. Passumpsic R.
contact metamorphic aure
temperatures raised to
sillimanite, andalusite zon
during intrusions of granit
magma
Sheffield
Stratford-Willoughby arch
24
Lyndonville
Victory pluton
23
Passumpsic valley esker;
longest, most continuous
of two-state region
Brattleboro syncline
Moose R.
sillimanite zone
22
Gardner anticline
21
Concord morai
20
19
St. Johnsbury
approx. west limit
Bronson Hill complex
sand/gravel pit with parabol
cross-section of esker
Knox Mountain pluton
Moore Dam
Comerford Dam
18
Barnet
Danville moraine
Ammonoosuc volcanics
Interstate 91
Barnet — Derby Line

east branch gives the ridge a total length of 24 miles, making it by far the longest, and certainly one of the best, eskers in Vermont and New Hampshire. It is on the west side of the river south of St. Johnsbury, where it crosses to the east side and remains there to Lyndonville and beyond. The best exposure is a sand pit where exit 20 north intersects US 5. From there, US 5 climbs to the top of the esker ridge and follows it through the village as Main Street. The ridge crosses the Passumpsic at the north end of Main Street. It is highest and most massive in the first 3 miles north of the crossing; there, however, it is partly buried under lake sediments deposited in a branch of Lake Hitchcock that reached nearly as far up the Passumpsic as the north end of the esker. The clean, clay-free materials of eskers are excellent for concrete and asphalt aggregates, roadbeds, and other construction uses. Several sand and gravel pits operate between St. Johnsbury and Lyndonville.

Most of the way between Lyndonville (exit 23) and Barton (exit 25—19 miles) in the beautiful valley of Miller Run, the highway follows another important glacial feature, the Lyndonville moraine. It connects with the larger north-south trending Danville moraine 12 miles northwest of Lyndonville. The moraine is pockmarked with sand and gravel pits in the first seven miles from Lyndonville. The abundance of moraines in the St. Johnsbury region, in contrast to their absence elsewhere, might be explained by the more subdued topography. Moraines are deposits of the glacier margin that are usually thickest and best preserved in the valleys, and thin and poorly preserved on the hilltops where they erode more rapidly. They tend to be more continuous and extensive in flatter country.

Excellent roadcuts expose the Waits River formation in the 10 miles northwest of Lyndonville. They show the thick marble layers, original limestones, that distinguish this unit from the Gile Moun-

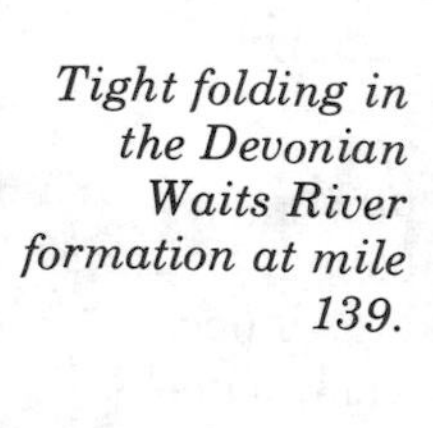

Tight folding in the Devonian Waits River formation at mile 139.

Thin, even bedding in Waits River formation slates and phyllites, mile 148.

tain formation. The marble is interbedded and intergraded with phyllites and schists, locally folded into rather fluid and streaky forms. The darker interlayers commonly contain innumerable thin lenses, layers, and "squiggles" of white quartz.

Northeast of Barton is Barton Mountain (2235 feet), a bare rocky hill carved from the granite of the Willoughby pluton of the New Hampshire series. Schists in the vicinity contain the minerals sillimanite and andalusite which crystallized under high temperature metamorphic conditions associated with intrusion of the granite. In fact, these minerals are common in all of the schists exposed along Interstate 91 between Barton and Derby Line (exit 29—22 miles), despite the distance of most of the rocks from exposed granite contacts. Heat emanating from igneous intrusions does not normally extend far into the enclosing country rocks, so more granite evidently lies at shallow depth. In fact, the many plutons in this region are very likely interconnected just below the surface, to form an enormous single batholith.

The road nips the west edge of the Derby pluton between exits 27 and 28 to Newport and exposes several characteristic igneous intrusive features. Contacts between the granite and the abundant blocks of schist engulfed in it are very sharp, except where the melt worked in between schist layers. The granite contains relatively large float-

ing crystals of white feldspar. Whitish granitic dikes that cut through both the schist and granite formed from residual melt that worked its way into fractures in the already crystallized granite near the margins of the pluton.

Glacial features abound between Barton and Derby Line. This was the site of the largest high-level meltwater lake in Vermont—Lake Memphremagog—during Wisconsin deglaciation. At its maximum, the lake extended arms into branching valleys, including those of the Barton, Black, Clyde and Missisquoi rivers. Consequently, this region is replete with lakeshore features and lake sediments, some high up on the valley sides. The much smaller and shallower modern Lake Memphremagog and many smaller lakes in this region are remnants of the original glacial lake.

Lustrous cleavage surfaces of phyllites in Waits River formation just past mile 161.

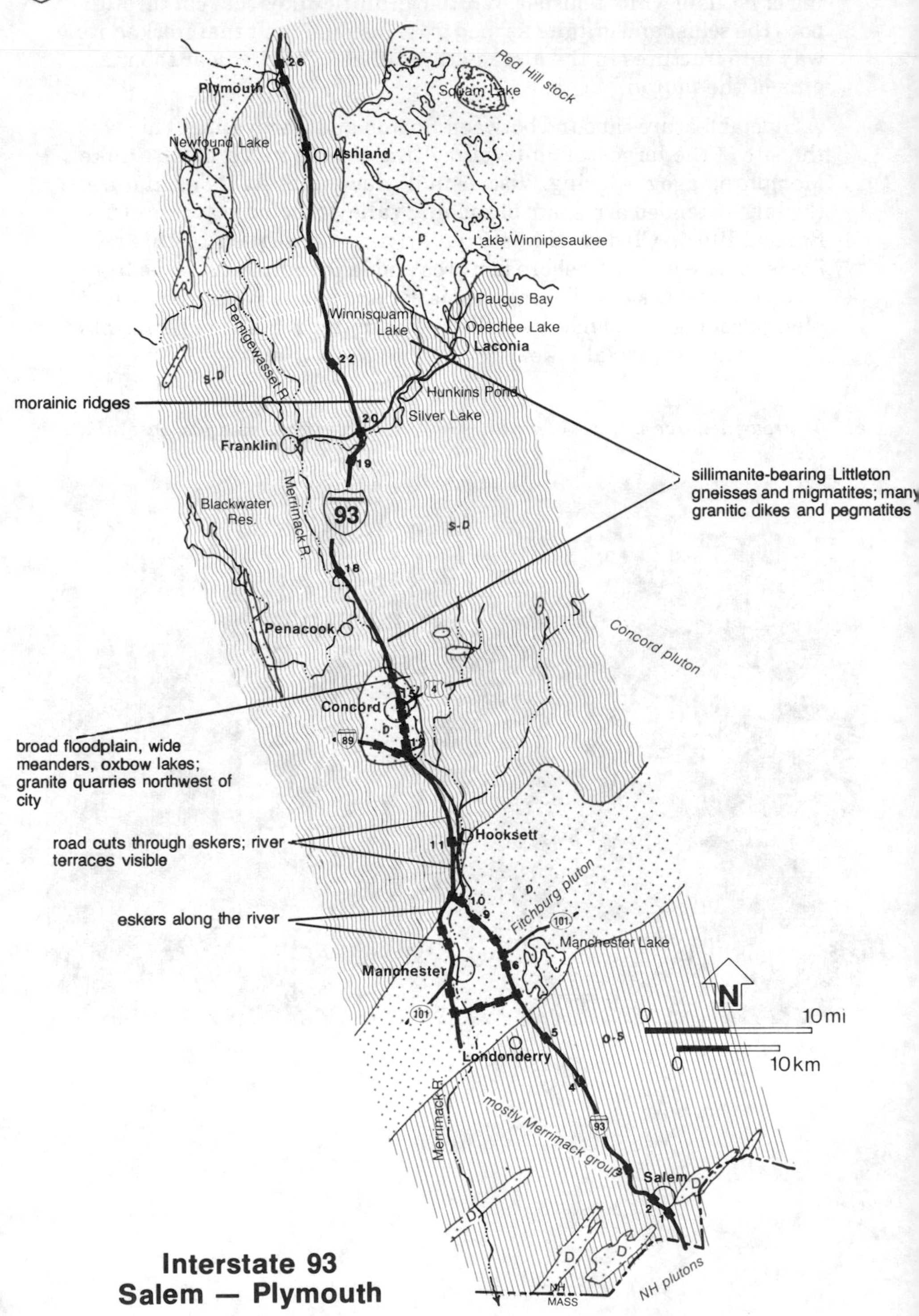

Interstate 93
Salem — Plymouth

Folded quartz-banded phyllite in early Devonian Littleton formation, 6 miles north of Massachusetts border.

Interstate 93: Salem—Plymouth, N. H.

75 mi. / 120 km.

Between Salem and Manchester (14 miles), Interstate 93 crosses the seaboard section of New Hampshire with its low relief and gentle southeast slope. The bedrock amply exposed in many roadcuts consists of strongly-layered schists and gneisses of the Merrimack group, of probable Ordovician-Silurian age. Most are products of medium-grade metamorphism in the biotite zone. Northwest of exit 5 to North Londonderry, 15 miles from Salem, the grade increases to sillimanite zone in just 2 miles near the contact with the Fitchburg pluton. Higher temperatures near the igneous intrusion raised the metamorphic grade. Manchester is right in the middle of the pluton. You pass diagonally through it in the circuit around Manchester and reach the opposite side at exit 11 to Hooksett. The many roadcuts are in pinkish granite or granitic gneiss cut by numerous pegmatites.

The route follows the Merrimack Valley from Manchester to Lincoln, but beyond Franklin the river becomes the Pemigewasset. It heads in the White Mountains on either side of Franconia Notch. The valley as far upstream as Plymouth is the site of former glacial Lake

Merrimack. South of Manchester, the lake was apparently made brackish by mixture with Atlantic waters as indicated by marine shells in the lake sediments. The lake lengthened gradually as Wisconsin ice receded northward during the waning stages of the Ice Age, and the meltwater took its place. The process lasted many centuries; varve studies indicate it took 500 years for the ice to recede just 20 miles from Concord to Franklin. Lake Merrimack drained to the sea much later as glacial rebound tilted the land surface southeastward.

This late glacial history is nearly identical to that of the Connecticut Valley, and the geological results are similar: shoreline features and terraces on the valley sides and thick sediment deposits under the valley floor. Here, however, unlike the Connecticut Valley, there are also numerous, well-preserved eskers, mostly on the west side of the river between Manchester and Concord. Eskers are common products of downwasting or receding ice, in contrast to active glaciers. The eskers are prime sources of clean, water-washed sand and gravel used in multiple building applications. One fine esker with several gravel pits crosses under the northwest end of the Merrimack bridge, near exit 10 on the north side of Manchester. Several more cluster about a mile southwest of exit 10. Interstate 93 goes right through one a mile south and another less obvious one, north of exit 11.

Valley terraces are also exceedingly well-defined and open to view between exits 10 and 11, where the highway is close to the river. To see them even better, follow New Hampshire 3A for 12 miles between exits 10 and 12. The route first traverses one terrace past kettle holes, following the curve of the exit 10 esker, and then descends to the river floodplain in view of an enormous terrace-top sand and gravel pit across the river. Near exit 11 it skirts the base of a 110-foot high terrace scarp for about 2 miles with similar scarps in view on the opposite side of the river. Then it ascends to another terrace top; just north of the Interstate 93 rest area, it goes over the top of another fine esker past kettle holes and a small kettle pond.

Strongly sheared gneiss near Manchester shot through with granitic veins and pegmatite.

Most of the Littleton formation rocks exposed in many cuts between Hooksett and Concord are gneisses and migmatites with cross-cutting granitic dikes and white pegmatites. Granite quarries are in Rattlesnake Hill northwest of the capitol, and several more lie within a 10-mile radius. The Concord granite has remarkably uniform composition and texture that make it excellent for building, cemetery monuments, or sculpture.

The route through Concord is on the floodplain in an extremely broad section of valley where the river meanders like a gigantic snake. Near exit 15 are several abandoned meander loops, three of which contain oxbow lakes. Horseshoe Pond, the largest, is west of the road just north of the exit.

The route between Concord and Plymouth (42 miles) passes over more Littleton formation. Most of the cuts are in grayish, sillimanite-bearing gneiss, some with migmatitic granitic veining and fine-grained granitic dikes. Many cuts are rusty, due to oxidation of iron minerals such as pyrite. Note, also, that several cuts retain their glacial polish where soil has only recently been removed.

Between Concord and Plymouth, the highway crosses the Lake Winnipesaukee outlet at exit 29 to Franklin and Laconia. The drainage route extends northeast through Silver Lake, Hunkins Pond, Winnisquam and Opechee lakes, and Paugus Bay, a route marked in several places by morainal ridges. Glacial Lake Winnipesaukee preceded the modern lake and occupied a much larger area. The lake enlarged gradually as the ice front receded northwestward uncovering the lowland basin. In the early stages, the lake outlet was southeast through Alton Bay into the Cocheco River and on to the Atlantic. The ice then stood approximately over the present outlet, blocking the way and leaving the moraines. Further retreat opened this outlet, permitting the lake to drop. Postglacial uplift has since reduced the lake to its present size.

Lake Winnipesaukee occupies a huge lowland underlain by coarse granite that is so crumbly as to be scarcely fit for use in road construction. Apparently the rotten rock was bulldozed by advancing ice sheets during the Ice Age, creating the lake basin. The surrounding hills are all composed of more resistant rock.

The highway is within view of the widely meandering Merrimack River for only 10 miles north of Concord, then swings away from it to New Hampton (exit 23, 31 miles from Concord). Between New Hampton and Plymouth (exit 25, 11 miles), the road follows the Pemigewasset River. Terracing here is lower and less conspicuous than that of the Merrimack Valley because Plymouth is at the northern limit of Lake Merrimack; the valley fill here was never deep.

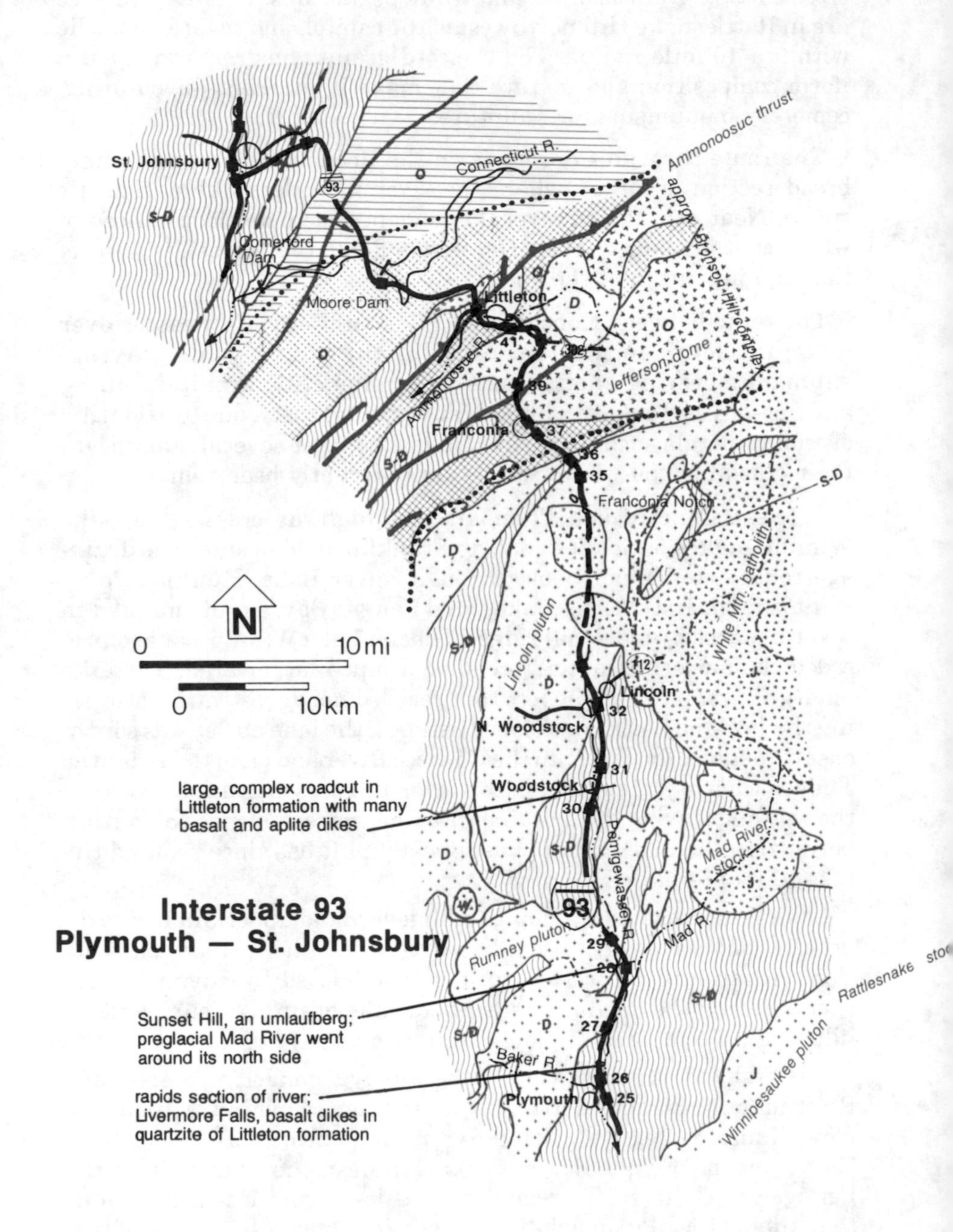

Interstate 93
Plymouth — St. Johnsbury

Interstate 93: Plymouth, N. H.— St. Johnsbury, Vt.

60 mi. / 96 km.

Plymouth is at the juncture of Baker River with the main branch of the Pemigewasset River. Interstate 93 follows the Pemigewasset River north to Franconia Notch (31 miles). The section of the main branch of the river between exits 26 and 27 (3 miles) is unique in having many rapids and no floodplain, indicating that, even today, it is actively cutting downward. One of the rapids sections, called Livermore Falls, is visible at close range from US 3 at little more than a mile north of exit 26. Well-exposed quartzite layers of the Littleton schist that dip about 60 degrees downstream hold the lip of the falls. Several dark green to black basalt dikes ranging from 2 to 10 feet thick cut these rocks. A broad flat shelf about 100 feet above the east side of the falls, complete with an arcuate meander scar, marks an earlier, higher course of the river. Referred to as The Plain, the terrace is more than half a mile wide in places and extends northward for three miles.

The section of valley near exit 28 is very broad where Mad River comes in from the northeast, and the Pemigewasset River meanders widely. Rounded Sunset Hill immediately north of the exit is separated from the hills north of it by a wide, sand-filled valley that appears to be the preglacial route of the Mad River to its juncture with the Pemigewasset River. The river now flows around the southeast side of the knob through a narrow rocky channel. These features suggest that during Wisconsin deglaciation the ice margin stood for a time more or less aligned with the present course of Mad River, blocking the old route north of Sunset Hill. Meltwater flowing

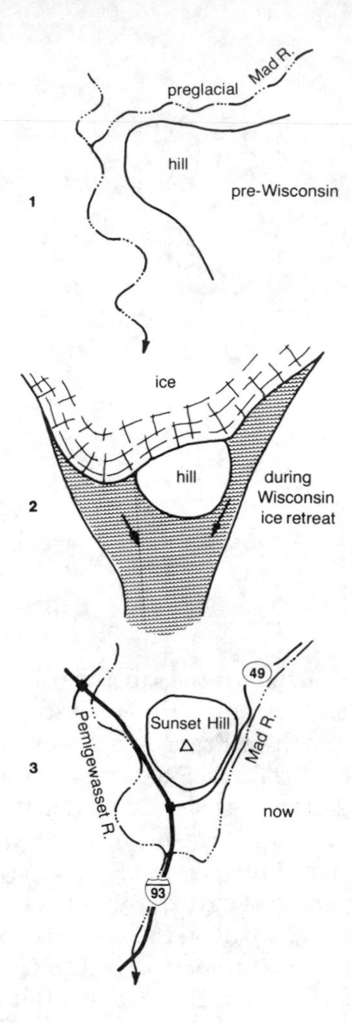

Theoretical sequence of events leading to formation of Sunset Hill, an apparent "umlaufberg."

through the valley in front of the ice had only one escape—through a bedrock pass that then connected the hill to the mountain east of it. By the time the old northern course was free of ice, the new channel was entrenched where it still remains several thousand years later. Much of the sediment fill in the broad valley below this may date from this late-glacial event.

Roadcuts between exits 26 and 28 are all in Littleton formation, rocks so intensely metamorphosed that they are difficult to distinguish from igneous rocks of the several New Hampshire series plutons in this region. Most exposures are criss-crossed by dikes of white aplite, a sugary-textured form of granite, and white pegmatite, both offshoots of the igneous plutons. These also occur within the plutons themselves, especially near their contacts with the country rocks. A

The Flume, a narrow slot where a basaltic dike has been removed by Flume Brook.

large body of intruding magma generally cools and hardens from the margins inward, so that residual melt pools in the core. Dikes form where this melt forces its way along cracks in the hardened margins and country rocks and there crystallizes. The complexity of the Littleton formation in this region is well-illustrated in the enormous cut between exits 30 and 31 on the east side of the river across from Woodstock. The host rock of strongly layered biotite gneiss is shot through with aplite and dark basalt dikes.

At Lincoln (exit 32), the Pemigewasset River divides into east and west branches. This is also at the intersection with New Hampshire 112, one of the most scenic mountain roads in the state. Interstate 93 continues north along the west branch of the river, which rapidly diminishes upstream to a mountain brook. Between Lincoln and Franconia (exit 37—13 miles), you pass through lovely Franconia Notch (1896 feet) under the watchful eyes of the Old Man of the Mountain. Several geologic features in the area are worth a pause and an easy hike. Most are maintained by the New Hampshire State Park Service, which preserves their natural setting despite hundreds of thousands of visitors each year.

The Flume, for example, is a beautiful, narrow gorge in the Conway granite near the base of Mount Liberty (4460 feet) 4 miles south of the Notch. The canyon, only 12-20 feet wide by 700 feet long and with vertical walls up to 70 feet high, formed as Flume Brook eroded a basalt dike. There is a 45-foot waterfall at the upper end of the gorge.

The Basin, a little more than a mile north of the Flume, is an enormous pothole that the Pemigewasset River carved into Conway granite. The abrasive action of sand and gravel in swirling eddies, especially during times of flood, excavates potholes. Some geologists estimate that the Basin required 25,000 years to achieve its present size. This, and many other potholes and water-polished surfaces in the Notch area, give credence to the Indian name, Pemigewasset, which means rapidly moving.

You can view the profile of the Old Man of the Mountain, made famous by Hawthorne's classic, "The Great Stone Face," from the north end of Profile Lake. The profile changes quickly as you move away from this vantage point. Overhanging slabs of granite near the top of the east face of Cannon Mountain form the profile of the Old Man. The cliff below the face is itself an interesting phenomenon. Since the Ice Age, it has been peeling like an onion, a process called exfoliation, that is common in massive granites. The same mechanism created Half Dome in Yosemite National Park and Stone

View south from Artists Bluff to Franconia Notch. Profile is definitely rounded as result of glacial scour although much is concealed by forest cover; Echo Lake in foreground.

The Basin, a large pothole carved by the Pemigewasset River.

Mountain of Georgia. A gigantic talus slope at the base of the cliff formed as broken slabs of granite peeled one by one from the face and tumbled. The cliff face has flights of upside-down steps marking where the slabs—the onion layers—broke off.

The best panoramic view of the Notch is from the top of Artists Bluff, a 500-foot high rock dome just north of Echo Lake. The setting resembles that of the north end of Crawford Notch farther east. As in Crawford Notch, the valley is glacially gouged and trends from north to south, with high mountains flanking the north entrance. Ice-rounded Artists Bluff blocks the entrance just as Willard Mountain does at the Gate of Crawford Notch. The glacial history, therefore, is probably similar. The Wisconsin ice sheet piled up against the mountain front and bulged into the Notch. At a later stage when the ice was thick enough to overtop the mountains, the Franconia Notch tongue at its base continued to move independently along the valley floor.

Cannon Mountain ski area is in full view from Artists Bluff. In the warm season, you can ride the aerial tramway to the summit for breathtaking panoramas of the surrounding countryside. Looking east, you can see all of Franconia Ridge with Mounts Lafayette (northernmost, 5249 feet), Lincoln (5108 feet), and Liberty (4460 feet). The summit ridge is scalloped by four bowl-shaped rock amphitheaters that look suspiciously like glacial cirques, at the heads of

gouged valleys that resemble small-scale versions of Franconia Notch. The steep headwalls of these valleys are especially prone to landslides. Major slides occurred on Mount Lafayette in 1826, 1850, 1883, 1915, 1947, 1948, 1959, and 1974!

Between Franconia and the Connecticut River (16 miles), Interstate 93 crosses the complex of metamorphosed volcanic and plutonic igneous rocks that comprise the Bronson Hill anticline, one of the most important elements in the plate tectonics interpretation. This is the collapsed volcanic archipelago formed early and shoved against the continent late in the Taconian mountain-building event as the proto-Atlantic Ocean began to close. The whole mass was further compressed between Africa and North America in Acadian time when the ocean basin clamped shut. The Taconian compression initiated the great westward-directed thrust faults of eastern Vermont.

Between Franconia and Littleton (9 miles), you cross the southwesten tip of the Jefferson dome, the largest of the Oliverian plutons that core the Bronson Hill anticline. This body continues northeastward for about 40 miles to the Maine border. A few roadcuts reveal the rock as gray gneiss invaded locally by dark basalt dikes.

Inverted steps on the east face of Cannon Mountain formed by exfoliation of the type that forms domes.

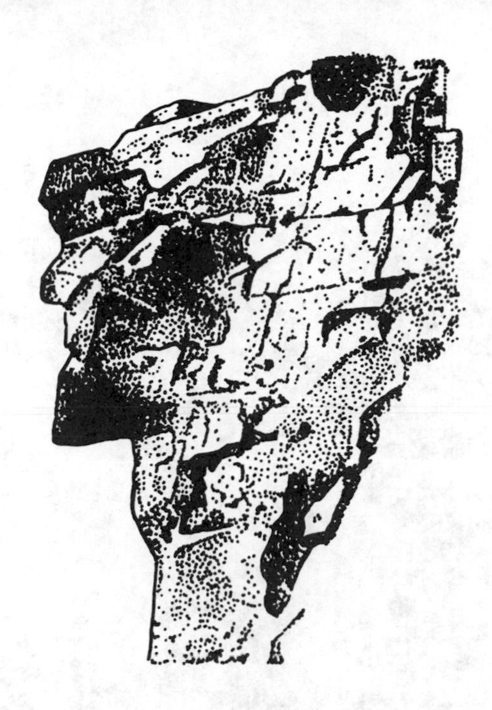

The Old Man of the Mountain on Cannon Mountain, as seen from Profile Lake. Face is about 40 feet high.

Between Littleton and the Connecticut River (8 miles), the route crosses major synclines and anticlines that trend northeast, and fault zones that trend northeast and dip northwest. They include the Ammonoosuc thrust fault that separates medium-grade schists containing staurolite on the east from low-grade greenschists of the Ammonoosuc formation formed from basalts on the west. The Connecticut River marks the approximate western limit of the Bronson Hill complex. Between the Connecticut River and St. Johnsbury, Vermont, (9 miles), you first cross the low-grade slates, phyllites, and micaceous quartzites of the Ordovician Albee formation exposed in many large cuts. Some cuts are rusty as a result of oxidation of iron-bearing minerals, and a few are shot through with aplite or rhyolite dikes, the latter a volcanic equivalent of granite. The last part of this traverse, between Stiles Pond and St. Johnsbury (4 miles), crosses Devonian Gile Mountain and Waits River formations, with schists amply exposed in large roadcuts, many full of thin, white quartz veins.

Dunham dolomite of the Champlain thrust slice near Interstate 89 junction with US 2.

US 2: The Champlain Islands

39 mi. / 62 km.

see map on page 50

The Alburg peninsula projects south into Lake Champlain from Canada between the Richelieu River, the lake outlet to the St. Lawrence, and Missisquoi Bay. The peninsula and the major Champlain Islands are like a great, low "stone fence" that divides the northern end of the lake into almost equal parts.

US 2 traverses the fence nearly from one end to the other, from the bridge to New York's Rouses Point at the Canadian border to the mainland causeway near Burlington. It is a scenic route through rolling farmland, blessed with open views across broad expanses of shimmering water to the shadowy forms of the Adirondacks on one side, the Green Mountains on the other. Pleasant villages with well-kept historic houses nestle in coves and bays dotted with small boats at anchor.

This peaceful island country rests upon bedrock of shale, limestone, and dolostone mostly belonging to middle Ordovician Trenton, Black River, and Chazy groups. It is relatively soft bedrock that, for unknown reasons, was less worn down by erosion than adjacent drowned parts of the Champlain lowlands presumably underlain by the same rocks. There are several reasons for the existence of the lowlands other than weak bedrock.

First, and perhaps most important, the Champlain lowlands are between the great west-directed thrust faults of western Vermont and the Adirondack dome of New York. The Adirondack margin, in particular, is cut by numerous block faults on which the Champlain side dropped. Second, it is the site of a preglacial drainage system that trended from north to south. Finally, this was a natural channel for ice erosion during the Wisconsin glaciation, presumably also during at least three preceding glaciations because it was in line with the general direction of ice movement. Geologists speak of the Champlain ice lobe that channeled its way south between the bordering highlands scouring, scraping, gouging, bulldozing, and scooping up materials, eventually to dump them somewhere farther south. It is certain that this erosion far exceeded that in the mountains, both in duration and intensity. The ice was here longest, and it was very thick when the ice sheets overtopped the mountains and reached as far as Long Island.

To fully understand the landscapes visible from US 2, you need also bear in mind the late-glacial history of this region. The Lake Vermont stage began when meltwater was trapped in the southern part of the lowlands in front of the receding ice, and it ended thousands of years later only after the Champlain lobe had completely wasted away and the ice blockage in the St. Lawrence Valley finally melted. Only then did Lake Vermont drain through the St. Lawrence to the sea. The lake level reached 1000 or more feet above the present level of Lake Champlain. A long period of erosion followed the draining of this great lake before the basin was again inundated, this time by Atlantic seawater, the Champlain Sea. Uplift of the land by glacial rebound eventually drained the brackish waters and left Lake Champlain as you see it today. All that left the bedrock surface of the islands, the lowlands bordering the lake, and the lake bottom deeply mantled with glacial drift and lake sediments, which add to the flatness of the terrain. The deposits were thicker and the surface flatter before the modern streams carved them up.

Strongly cleaved Stony Point shale on South Hero Island near village of South Hero. Note remarkable cleavage refraction, or change in direction, across the thick carbonate bed in middle of photo. Bedding is nearly horizontal, especially notable below carbonate bed.

Crumpled Stony Point shale in quarry on South Hero Island near causeway to mainland.

Trenton group limy black shales, revealed in one or two roadcuts near South Alburg, underlie all of the Alburg peninsula. A worthwhile side trip from there is by way of Vermont 129 to the St. Anne de Beaupre Shrine on Isle La Motte (5 miles from South Alburg). This is the site of French Fort St. Anne, Vermont's oldest white settlement, built in 1666 by Capt. Pierre La Motte for defense against the Mohawks. Samuel de Champlain had discovered the lake much earlier, in 1609, when he came up the Richelieu during his exploration of the St. Lawrence River. The statue of Champlain at the St. Anne site was carved from a massive block of granite from Barre, Vermont. The lakeshore at the shrine is typical of all the islands, cluttered with many erratic boulders of Precambrian gneiss that washed out of the glacial till. These were picked up and transported by the advancing

Middle Ordovician Trenton group limestone cliff on South Hero Island near ferry dock.

Wisconsin ice from the Canadian shield far to the north, then dumped here. It is also likely that the stony till, a direct ice deposit, was exhumed from beneath a thick cover of glacial outwash and lake sediments. A quarry on Isle La Motte in the Crown Point limestone produces a dark, highly fossiliferous, ornamental stone that polishes well.

Roadcuts near the village of North Hero expose more Trenton group limy black shale. Most of the old stone homes on the islands are made of lighter gray, more massive limestone, probably from the Chazy or Black River groups. Near the village of South Hero are large roadcuts in Trenton limy black shale in which bedding is nearly horizontal and moderately folded. Look for the strong vertical cleavage that transects and slightly offsets the beds. In one place, the cleavage sharply changes direction where it crosses an 8-inch-thick layer of limestone. In places, the black shale is laced with veins of white calcite precipitated from groundwater.

You cross to the mainland southeast of South Hero, and then pass through about 2 miles of marshy lowlands of Sandbar State Park. This is the modern delta of the Lamoille River, one of three major rivers that cut through the Green Mountains and empty into Lake Champlain. The even larger Winooski delta lies immediately south near Burlington; the Missisquoi delta is at the north end of the lake east of Alburg.

At the eastern edge of the park, you cross the Champlain thrust fault into hilly country with many large roadcuts. Most of the cuts are in buff to pink, locally mottled, massively bedded Dunham dolomite of early Cambrian age. Near the Interstate 89 interchange the formation contains light gray quartzite interbeds.

Early Cambrian quartzite possibly belonging to Potsdam sandstone south of US 2 / Interstate 89 junction

US 2: Montpelier, Vt.— Maine Border

95 mi. / 152 km.

Between Montpelier and Marshfield (16 miles), the road follows the Winooski River. The valley as far as Plainfield (9 miles) contains sediments and shoreline features that mark the easternmost reach of glacial Lake Vermont. Plainfield is more than 500 feet higher than modern Lake Champlain, and that gives some indication of the great size and depth of Lake Vermont. Up the valley from Plainfield, sediments and shorelines record an earlier, high-level glacial lake that reached as far as Cabot, five miles northeast of Marshfield.

Roadcuts between Montpelier and Plainfield expose mica schists and interbedded marbles of the Devonian Waits River and Gile Mountain formations. Midway between Plainfield and Marshfield (7 miles), the Waits River formation is in contact with granite of the Knox Mountain pluton, the largest of the New Hampshire series in Vermont. The road follows the contact for 8 miles almost to West Danville, past roadcuts and small quarries that illustrate some of the intrusive relationships: homogeneous granite invading between the layers of the schist; stray chunks of schist in the granite; and numerous aplite and pegmatite dikes in both the granite and the country rock. The schists increase in metamorphic grade between Montpelier and the margin of the pluton as a result of contact metamorphism associated with the intrusion. The closer to the intrusion the higher the temperature; some of the schists near the contact closely resemble the granite.

Between West Danville and Danville (3 miles), the road crosses the Danville moraine, one of the largest and best-developed in New England. Danville is about in the middle of the moraine belt, which

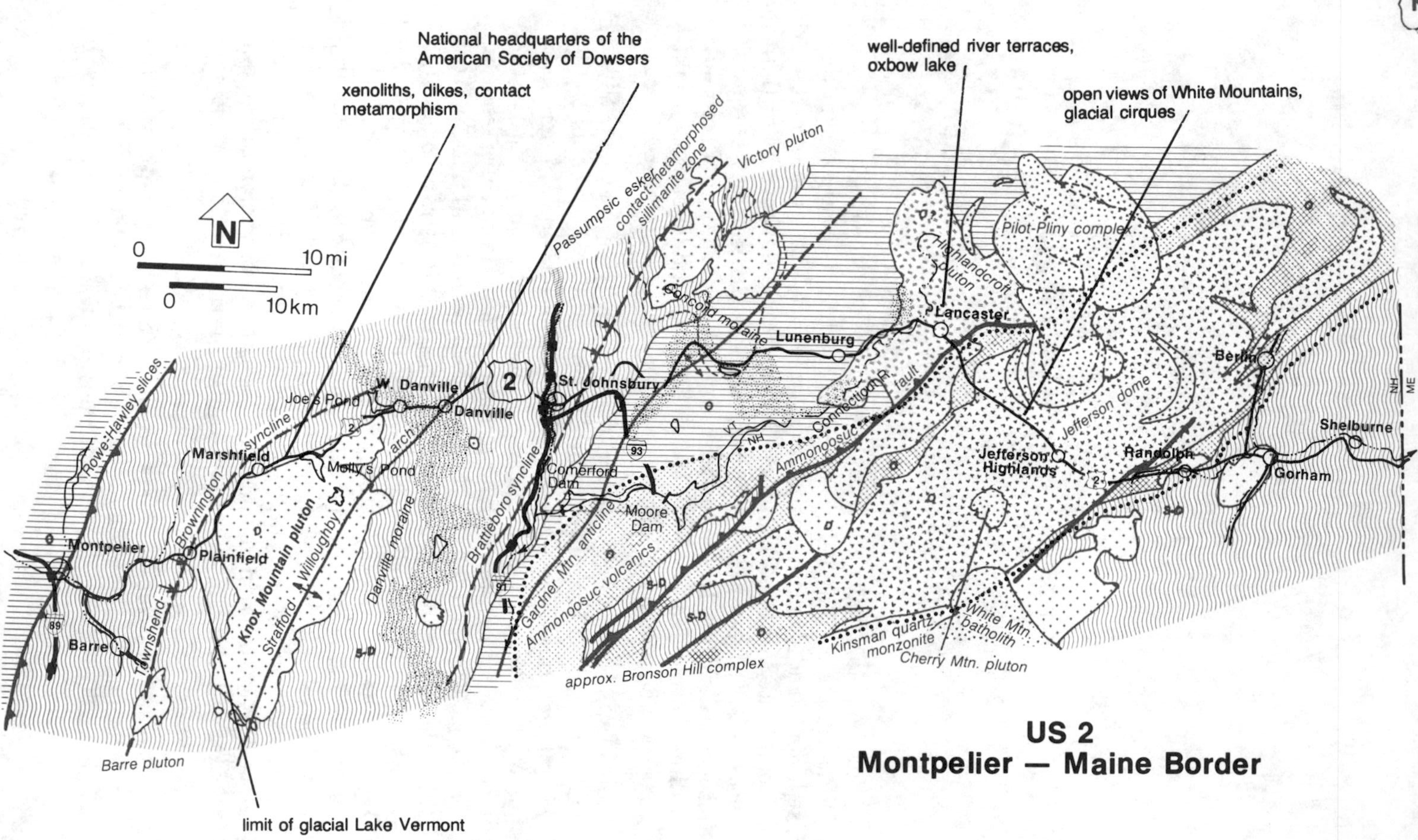
2
National headquarters of the American Society of Dowsers
xenoliths, dikes, contact metamorphism
well-defined river terraces, oxbow lake
open views of White Mountains, glacial cirques
limit of glacial Lake Vermont
N
0
10 mi
0
10 km
Rowe-Hawley slices
Montpelier
Barre
Barre pluton
Townshend-
Brownington
syncline
Plainfield
Marshfield
Joe's Pond
W. Danville
Danville
Molly's Pond
arch
Knox Mountain pluton
Stafford-Willoughby
Danville moraine
Brattleboro syncline
St. Johnsbury
Comerford Dam
Moore Dam
Passumpsic esker
contact-metamorphosed sillimanite zone
Victory pluton
Concord moraine
Lunenburg
Lancaster
Highlandcroft pluton
Pilot-Pliny complex
Gardner Mtn. anticline
Ammonoosuc volcanics
Ammonoosuc
Connecticut R.
fault
VT
NH
Jefferson dome
Jefferson Highlands
Randolph
Berlin
Gorham
Shelburne
NH
ME
approx. Bronson Hill complex
Kinsman quartz monzonite
White Mtn. batholith
Cherry Mtn. pluton
S-D
93
91
89
US 2
Montpelier — Maine Border

extends southward 30 miles to Goose Green, northward 20 miles to a point 2½ miles west of Glover. Recognize moraines by their hummocky topography, stony fields, and in many cases by the stone fences that divide farm fields. For unknown reasons, moraines are much more scarce in Vermont and New Hampshire than in western and southern New York.

Danville is the national headquarters of the American Society of Dowsers, a group that claims to locate water by using a forked stick or other device. In the traditional method, the dowser holds a forked stick horizontally in front of him with both hands and walks over the ground, locating where the stick supposedly points down to the ground of its own accord. Other methods employ a pendulum, angle rods, or wand or bobber of variable designs. The society in 1985 had a membership of over 3500 with 59 chapters in the United States. This, however, is probably only a fraction of the total number of practitioners in the country. I do not believe in dowsing as a valid method of locating groundwater, for it has no scientific basis; but I think it is here to stay. It is fairly safe to assume that there is groundwater in almost any part of Vermont and New Hampshire, which may explain why the dowsers report a high rate of success.

St. Johnsbury, 7 miles east of Danville, is in the Passumpsic Valley, site of the largest and most continuous esker in the region. You go right over the top of it in passing through the village on US 2. The oldest and most beautiful section of the town, called The Plain, is on the crest of the esker, with Main Street running northeast along its axis. US 2 descends the steep flank of the esker along Eastern Avenue. The materials and cross section of the esker are exposed in a sand and gravel pit at the south end of South Main Street. Several other quarries operate along its 24-mile length to the north.

Between St. Johnsbury and the Connecticut River (26 miles), the highway crosses the Concord moraine, revealed in till banks, stony fields, hummocky topography, and boggy wetlands. The moraine occupies a wedge-shaped region with the apex pointing north at the hamlet of Victory. US 2 follows it for 9 miles from Concord east. This is one of four large moraine belts of northeastern Vermont, the biggest of which is the Danville moraine discussed above; the others are the Lyndonville and Burke moraines northwest of the Concord moraine. Nobody knows why moraines are better developed in this region than in other parts of Vermont and New Hampshire, but it may have to do with the lower relief. Flatter terrain is more conducive to gradual retreat of ice, as opposed to downwasting and fragmentation; moraines mark temporary still-stands of the ice front.

East of Lunenberg, the road descends to the floodplain of the Con-

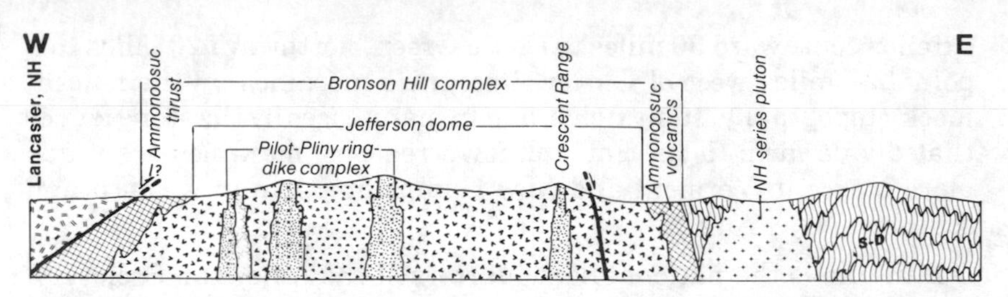

Schematic east-west cross section through Jefferson dome north of US 2.
—Adapted from M. P. Billings.

necticut River, then follows the widely meandering stream four miles to Lancaster, New Hampshire. At one point the road skirts a large oxbow lake, a cut-off meander loop. An exceptionally well-defined terrace at Lancaster was carved by postglacial downcutting of the river into valley-fill sediments.

Some of the most magnificent mountain scenery of New England lies along US 2 between Lancaster and Gorham (25 miles). The road trends southeast to Jefferson Highlands, and the great mass of the Presidential Range, with Mount Washington at its crest, dominates the view. The somewhat bald granite summits of the Pliny and Pilot ranges rise to the north, and the high Kinsman and Franconia mountains flank Franconia Notch to the south. Between Jefferson Highlands and Randolph (8 miles) are splendid grandstand views of the profound glacial cirques of the north end of the Presidential Range. It's quite a spectacle.

The route to Gorham is almost entirely over granitic gneisses of the Jefferson dome, the largest of the plutons that intrude the metamorphosed volcanic rocks of the Ordovician Ammonoosuc formation and form the core of the Bronson Hill anticline. Mount Waumbek, directly north of Jefferson Highlands, contains one of the many ring-dike complexes that comprise the White Mountain magma series of New Hampshire; this one intrudes like a plug through the Jefferson dome and contains a number of more or less circular granitic bodies. The Crescent Range directly north of Randolph contains one of the ring dikes several miles from the central core. Rocks along the road between Gorham and the Maine border are high-grade gneisses and schists of the Devonian Littleton formation shot through with aplite and white pegmatite dikes. The presence of the mineral sillimanite in the schists shows that they were metamorphosed at very high temperature.

White pegmatite of the New Hampshire plutonic series intruding Littleton schist near Shelburne.

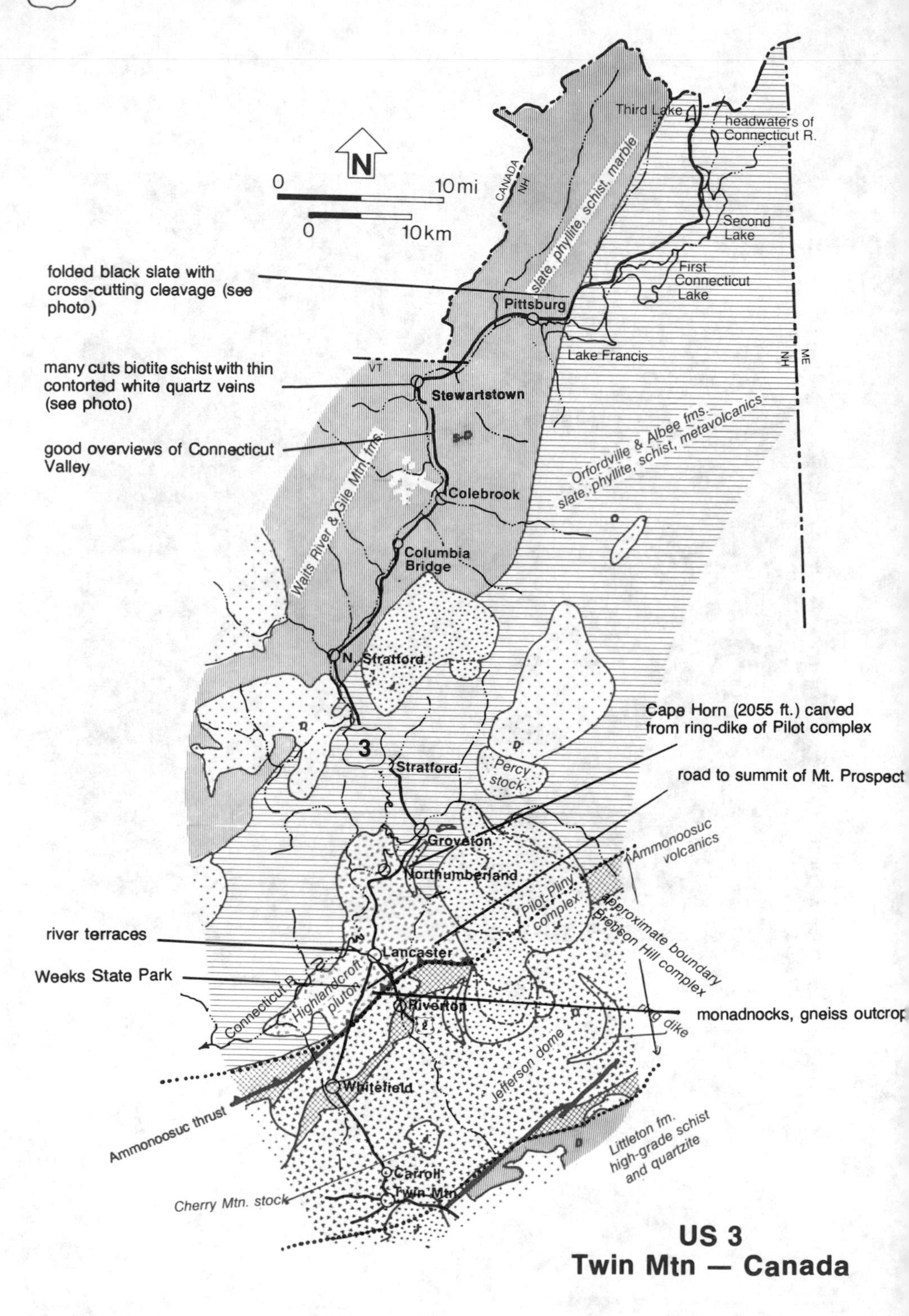

US 3
Twin Mtn — Canada

Highly deformed, dark grey phyllitic schist with many white quartz lenses, US 3, 37 miles south of Canadian border.

US 3: Twin Mountain, N. H.—Canadian Border

98 mi. / 157 km.

The Upper Connecticut Valley

Between Twin Mountain and Lancaster (17 miles), US 3 crosses Jefferson dome, the largest of the granitic plutons that core the Bronson Hill anticline. The domical mountain east of the road between Twin Mountain and Whitefield (8 miles) is the site of the Cherry Mountain stock, a small, circular body of granite of the White Mountain magma series that intrudes Jefferson dome. Halfway between Whitefield and Lancaster, you cross the Ammonoosuc thrust fault, out of the grayish to pinkish gneisses of the Jefferson dome and into the darker, coarser-grained rocks of a large Highlandcroft series pluton of probable Ordovician age. The rock is exposed in cuts on both sides of the road about 6 miles north of Whitefield, 3 miles south of Lancaster, where it consists of a medium-grained greenish granite mottled with pink potassium feldspar crystals. A little farther north is Weeks State Park where there are more outcrops on Prospect Mountain. A road leads to the summit of the mountain with its birdseye view of the surrounding country including the Connecticut Valley and distant Green Mountains to the west. The White Mountains spread out to the north, east, and south. Pleasant Hill, directly west on the other side of US 3, is held up by an island of resistant Albee formation slates and phyllites.

Lancaster is in a broad, flat section of the Connecticut Valley at the juncture of Israel River. Glacial Lake Hitchcock sediments that floor the valley are terraced as a result of postglacial downcutting of the river.

The prominent hills east of the road between Lancaster and Groveton (9 miles) are in the Pilot-Pliny complex of igneous stocks and ring dikes of the White Mountain magma series. Near Groveton, the road skirts the west slope of a slim arcuate hill called Cape Horn (2055 feet), carved from the outermost ring dike in this complex.

All of the bedrock along US 3 between Groveton and North Stratford (13 miles) is slate, phyllite, and schist of the Ordovician Albee formation and between North Stratford and Pittsburgh (32 miles) is slate, phyllite, schist and marble of the Devonian Gile Mountain formation. Between Pittsburgh and the Canadian border are more Orfordville rocks. The dark rocks locally contain thin, white, highly contorted quartz veins, and many display strong cleavage. Roadcuts are scarce.

Folded grey phyllites, US 3, 25 miles south of Canadian border. Strong, nearly vertical cleavage is transverse to the layering.

The most interesting geologic aspect of this route is that it follows the main branch of the Connecticut River to its source at the Connecticut Lakes. Glacial Lake Hitchcock reached at least to the northeast corner of Vermont, and the valley contains shoreline markings and terraces like those that are so well-developed farther south. Some are visible from the highway. This is a sparsely populated section of the state with a lot of till banks by the road and stony fields. The Connecticut Lakes are popular with fishermen and boaters.

Fields like this, cluttered with glacial erratics, supplied the materials for the many stone fences of parts of New England.

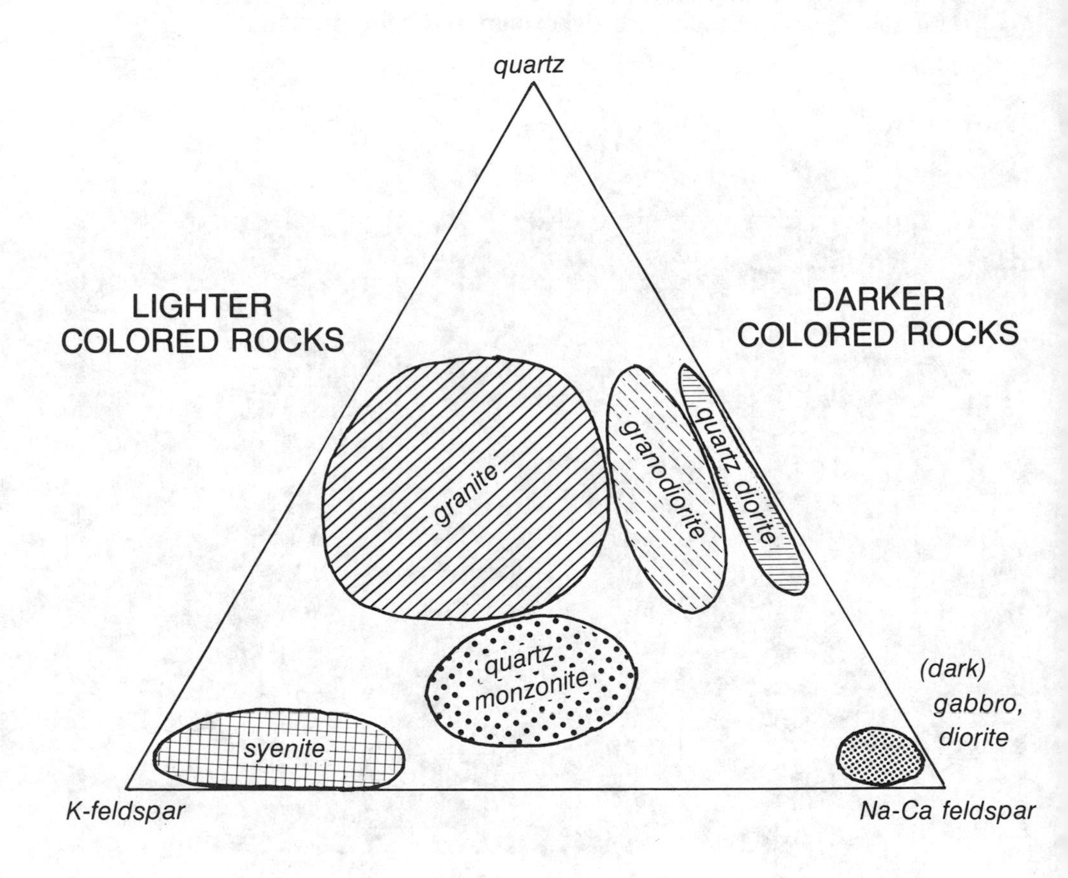

Approximate composition fields of major igneous rock types mentioned in the text, in relative abundances of quartz and the two kinds of feldspar, dark minerals ignored. Corners represent 100 percent of that mineral, and percentage decreases to zero across the triangle to opposite face. Granite at the exact center, for example, has 33 1/3 percent of each mineral.

US 4: Durham—Concord

36 mi. / 58 km.

see map page 202

Durham is in the core of the Exeter pluton, which intrudes Merrimack group slates and phyllites. The igneous rock is referred to as the Exeter diorite, but its variable composition averages to a somewhat lighter gray rock more akin to granite. In places, as at Durham, it is granitic whereas elsewhere, as at three miles northwest of Exeter, it is darker gabbro, the coarse-grained equivalent of volcanic basalt. Two miles southeast of Durham it is again very close to granite. Several roadcuts on US 4 near Durham demonstrate the range of composition and character of this igneous intrusion. Near its contact with the country rocks, a few cuts display a chaos of schist, gneiss, and even diorite blocks floating in an igneous matrix.

Between Durham and Northwood (15 miles), you pass from the chlorite zone to biotite to garnet to staurolite to sillimanite zones in rapid succession, and then cross into the Fitchburg pluton. The fast succession of metamorphic zones reflects a steep temperature gradient, no doubt because of heat from the igneous intrusion. One cut five miles east of Northwood reveals the Fitchburg pluton as a white granite cross-cut with white pegmatite dikes and full of large inclusions of coarse-grained gneiss with mica that sparkles in the sunlight.

Six miles west of Northwood, the road skirts Northwood Lake which straddles the western margin of the Fitchburg pluton. From here to Concord (15 miles) all of the bedrock is Littleton schist, but there are virtually no cuts. Near Concord are roadcuts in whitish,

medium-grained micaceous Concord granite containing a large mass of white pegmatite. The Concord granite is world-famous as building and monument stone.

Numerous lovely stone fences that stand in the region around the state capitol tell of the abundance of stony glacial till. Most were built during early settlement of the state, as the fields were cleared for farming. They are still beautifully kept today, reminders of Robert Frost's ironic observation that "Good fences make good neighbors."

The till deposits are mostly ground moraine, material deposited either under the active Wisconsin ice sheet or released directly from the stagnant ice as it melted downward. For unknown reasons, ridge-like end moraines marking prolonged stands of the ice margin are rare in New Hampshire.

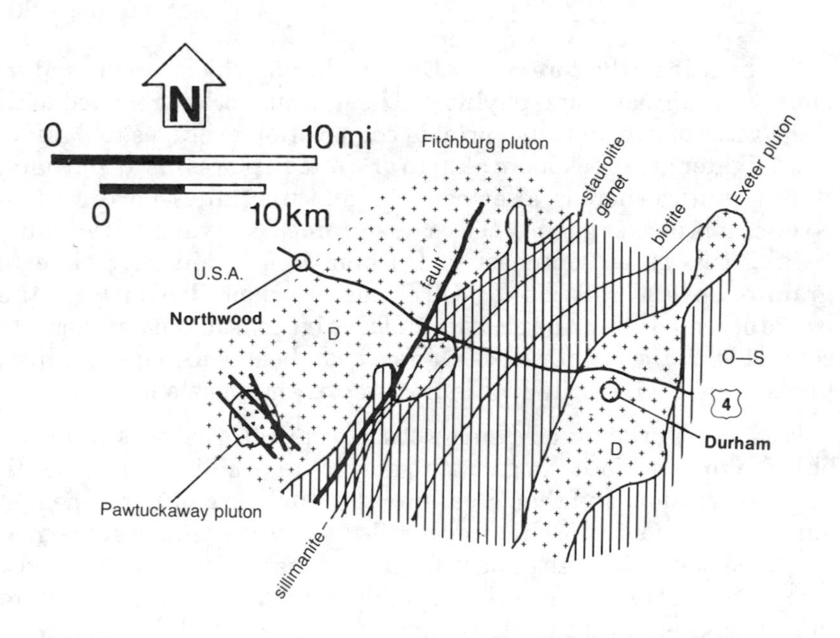

Lines labeled "biotite," "garnet," etc., mark first appearance of these metamorphic index minerals when traversing east to west, indicating increasing temperature of metamorphism approaching the Fitchburg pluton, an example of contact metamorphism.

Quechee Gorge in Ottauquechee formation was excavated by rapid outflow of proglacial Lake Ottauquechee.

US 4: White River Junction— Fair Haven, Vt.

64 mi. / 102 km.

Between White River Junction and Woodstock (14 miles), US 4 crosses the Brattleboro syncline over micaceous greenschist and quartzite of the Devonian Gile Mountain formation. The syncline tilts to the west so both limbs, or sides, of the fold dip eastward, or are overturned. The rocks are well-exposed in Quechee Gorge, six miles from White River Junction and often referred to as Vermont's Little Grand Canyon. The rock layers there dip steeply into the east wall of the gorge forming overhanging ledges and inclined slabs that face the west wall. The Quechee Gorge is about 165 feet deep and a mile long. Torrential outflow from glacial Lake Ottauquechee, a high-level meltwater lake that existed farther west in the Ottauquechee Valley, carved it. Many such lakes in Vermont were banked behind ice and glacial debris as the Wisconsin ice sheet wasted over the state's mountainous terrain.

The original bridge spanning the gorge was built in 1875 for the

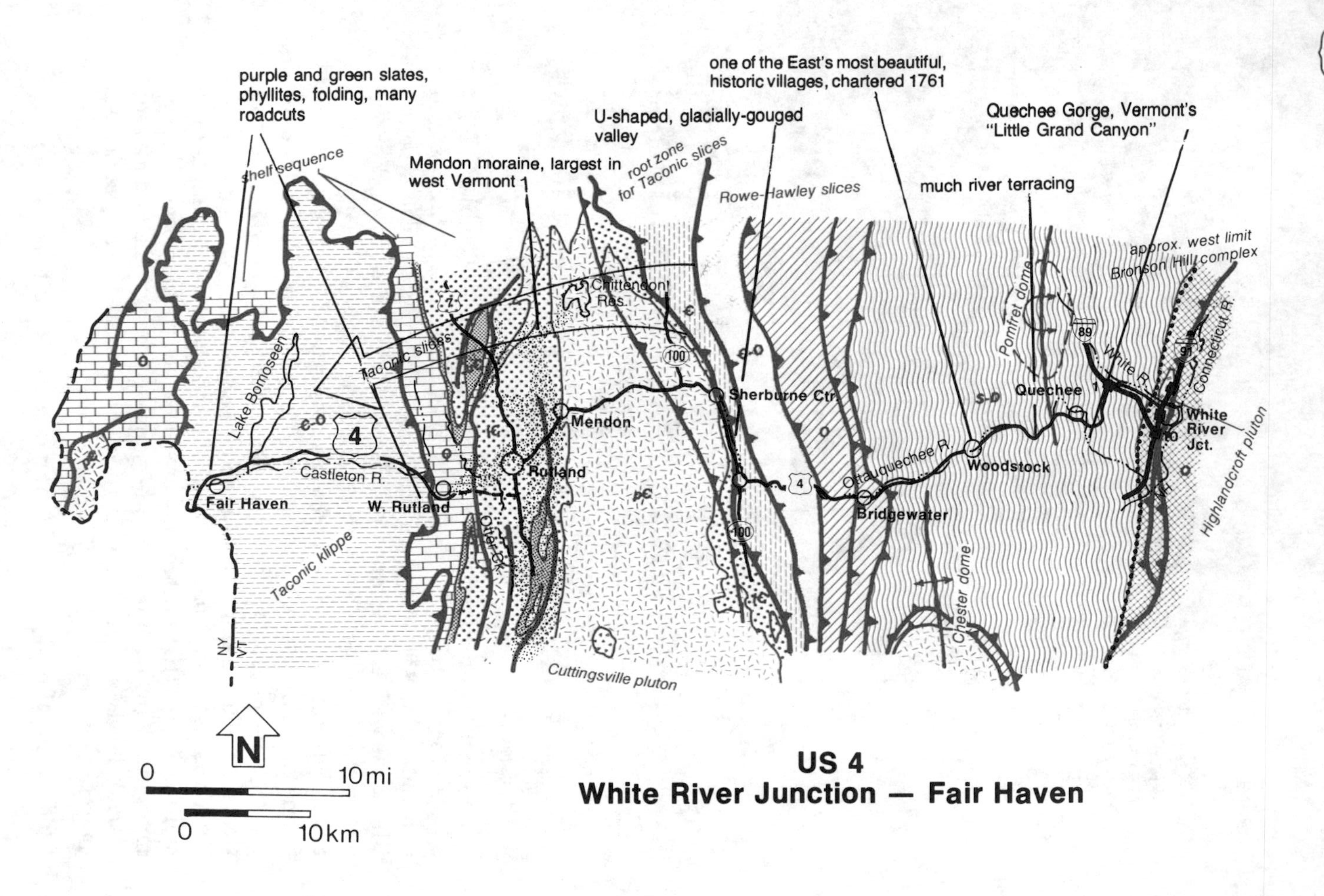

US 4
White River Junction — Fair Haven

Woodstock Railroad. The steel arch bridge used today, built in 1911, also served the railroad but was converted to automobile use when the railroad was dismantled in 1933. US 4 now approximately follows the old railroad grade. The valley between Quechee Gorge and Woodstock is conspicuously terraced and floored with sandy sediments that have been widely exploited judging from the numerous quarries, or borrow pits, visible from the road.

Woodstock, chartered in 1761, is one of the most charming villages in New England, retaining the physical and spiritual flavor of an earlier day.

Bedrock between Woodstock and Bridgewater (6 miles) is Devonian Waits River formation, consisting of schist, quartzite, and marble; but it is poorly exposed except in places along the river bed. You cross a major boundary just east of Bridgewater; from there to West Bridgewater (8 miles) the rocks are metamorphosed Ordovician and Cambrian sedimentary and volcanic rocks that flank the domical Precambrian core of the southern Green Mountains. You also cross the serpentine belt where there are numerous slivers of serpentine and a lot of minerals of interest to collectors. This is a zone of intense shearing and thrust faulting, reflected in the shattered, crumpled state of the schists exposed in roadcuts.

Between West Bridgewater and Sherburne Center (4 miles), US 4 goes north along major thrust faults near the Precambrian boundary. Rocks within the thrust fault slices, mainly schists of the Cambrian Camels Hump group, moved up and over the Precambrian massif and later eroded from its crest. The enormous Taconic klippe is a remnant of the same thrust slices on the other side of the mountains 15 miles to the west. Here, the Ottauquechee River has positioned itself in the

Covered bridge over Ottauquechee River at Woodstock.

fault zone. The broad valley has the characteristic gouged form indicative of ice erosion.

Between Sherburne Center and Mendon (9 miles), the road goes over the hump through Sherburne Pass from one side of the Precambrian massif to the other. Almost no rock is exposed along the road. Sherburne Pass is one of several major mountain gaps in the Green Mountains that were reshaped by ice. It is in a class with Smugglers and Hazens notches, and Lincoln, Middlebury, Mount Holly, and Brandon gaps.

Mendon stands on the Mendon moraine, the largest and most continuous moraine belt of western Vermont. Here and along the Green Mountain slope north and south up to an elevatiion of 1500 feet, it is really kame-moraine, meaning clean, water-washed sand and gravel deposited in standing water along the ice margin, but with a hummocky morainic topography. These deposits have great economic value as sources of construction aggregate, and support numerous quarries. True moraine makes up the bulk of the Mendon moraine on the floor of the Vermont Valley. It is composed of sediments released directly from the ice as it melted and contains more of the clay and silt that make it less suitable for construction.

Between Mendon and West Rutland (9 miles), you cross the beautiful Vermont Valley that separates the Green Mountains from the Taconic Mountains. The valley is carved from bands of Cambrian and Ordovician marbles and interbedded quartzites, phyllites, and other rocks that trend from north to south and collectively offered less resistance to erosion than the bedrock of the flanking mountains. The abundant marble is the basis for the Vermont marble industry, now centered northwest of Rutland, in Proctor.

In passing through Rutland, the road descends to the level of Otter Creek then follows the creek and tributaries to West Rutland. A little farther along, the road crosses a low divide, then follows Castleton River all the way to the New York border.

West Rutland lies almost astride the profound geologic boundary of the basal thrust fault of the Taconic klippe. Overall, the klippe is

Marble wall at Rutland.

Early Cambrian St. Catherine formation slate of the Taconic klippe, west of Rutland. Layering is gently inclined to right and cleavage is steeply inclined to right.

about 165 miles long from north to south by about 10-20 miles wide. Many small satellite klippen have been erosionally isolated from it.

The predominant carbonate rocks of the Vermont Valley, which presumably continue under the klippe, are metamorphosed shallow marine limestones, dolostones and subordinate shales and sandstones. They were deposited on the continental shelf of ancestral North America before the Taconian mountain-building event. Those rocks are deformed, but essentially in place. They have not been shoved very far westward along thrust faults. Original rocks of the klippe were mostly shales, probably deposited farther east at the edge of the continental shelf. Thrust-faulting then shoved the shales up and over the shelf sediments. They were highly folded and cleaved, and subjected to low-grade metamorphism as they were carried many miles westward. What you see today, though huge, is only a small part of the original thrust slices.

Rocks of the Taconic klippe are well-exposed in fresh roadcuts along the new scenic highway between West Rutland and the New York border (15 miles). Most of them are dark slates or phyllites with thin white quartzite interbeds or veins, or metamorphosed muddy sandstones. All belong to the Cambrian St. Catherine formation.

Fair Haven, the center of Vermont's slate industry, lies at the north end of the Vermont and New York slate district, an area about 8 to 10 miles wide from east to west and 25 miles long from north to south, that straddles the state line. The district is the only one in the country that produces green, purple, mottled, and red slate. There are also

Folded and cleaved St. Catherine formation slate.

black slates, but only small amounts are quarried. New York contains all of the red slate. There are many colorful green, purple and mottled slate roadcuts along US 4 between West Rutland and Fair Haven. More than 100 slate quarries exist in the Fair Haven, Castleton, Poultney, and Pawlet areas, most now abandoned. Slate production is an old industry now in its declining years, suffering from high wastage, costly production, and competition from synthetic substitutes. Today, most of the slate is used for roofing, ornamental floor tile, dimension stone, blackboards, flagging, billiard table bases, and in crushed form, for roofing granules, asphalt roofing tile aggregate, or other crushed stone applications.

The discovery of slate in this district dates back to 1839. At that time, there was little use for the stone locally, no convenient way to transport it. Railroads came in 1851, and two Welsh families organized the first slate company in Middle Granville in 1852, the Penrhyn Slate Company. As the industry grew, more slate miners came to the region from Wales to work in the quarries and mills. Descendants of these people are, in large part, the owners and operators of today's industry.

Quarrying and processing slate for use as floor tile illuminates some of the physical and geological characteristics of the stone. Quarrymen remove thick slabs by drilling one-inch holes spaced three inches apart along the cleavage, then blasting with black powder.

They pry the loosened slabs apart with crowbars and sledge hammers. Select pieces, averaging only about 30% of the total, go to the mill for processing; the rest is waste. At the mill, the slate may be further split with a jackhammer and cut into desired dimensions with a diamond saw. For floor tile, slabs are cut to specific square and rectangular shapes. Splitters in the mill then cleave these dimensioned slabs into sheets slightly thicker than ¼ inch, using a wide chisel and a hammer. About half of the sheets are discarded for irregularities. The stone will split at almost any place where the chisel blade is placed parallel to the cleavage, and the ease with which it splits is quite remarkable considering that the cleavage planes are generally invisible.

The final step is gauging. The split sheets ride a conveyor belt through a horizontal milling machine that shaves them to a uniform quarter-inch thickness. The milled surface permits the tile to be laid with an all-purpose cement like vinyl floorings. The tile is commonly sold in cartons containing mixed colors and sizes that can be set in various mosaic patterns.

Total wastage from quarry to final cut-stone products of floor and roofing tile and dimension stone averages 95 percent in Vermont, 85 percent in New York. Survival of this portion of the slate industry hinges on reducing wastage.

Dot-dash pattern of quartz lenses in St. Catherine formation slate.

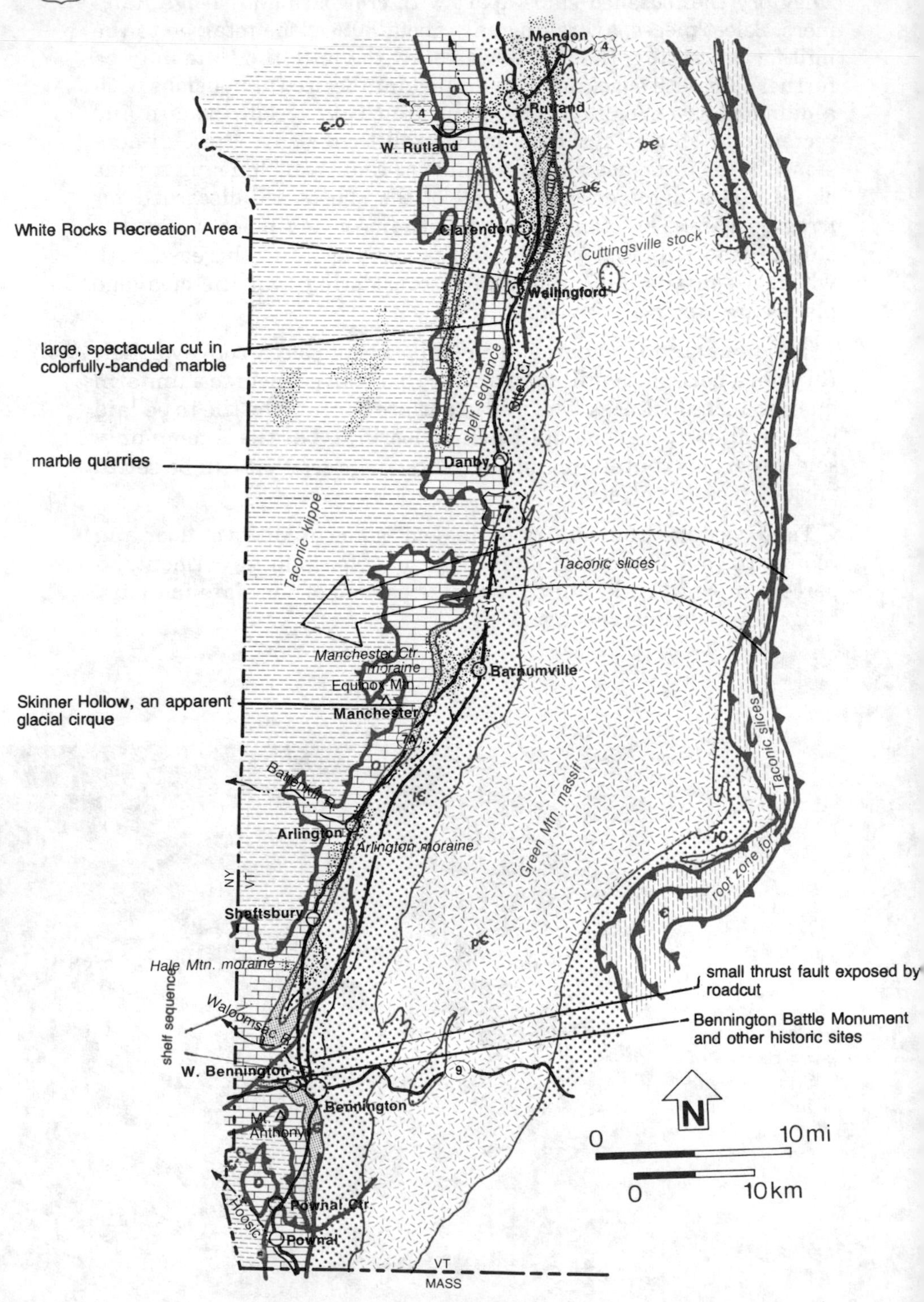

US 7
Massachusetts Border — Rutland

US 7: Massachusetts Border—Rutland, Vt.

66 mi. / 106 km.

The Vermont Valley

Roadcuts between the Massachusetts border and Pownal Center (5 miles) expose brown to black mica schists of the Ordovician Hortonville formation. The schists are interlayered with light gray marbles and both are much folded. Deformation here is at the base of the Taconic klippe, the mass of rock that was shoved many miles westward from its place of origin during Taconian and Acadian mountain-building, and then isolated by erosion. Pownall Center is in a segment of the klippe made up of phyllites and purple and green slates of the Cambrian St. Catherine formation. Note that these rocks are older than those under the basal thrust fault, a common condition in many parts of western Vermont.

The segment of US 7 between Pownall and Pownall Center (3 miles) rises 400 feet across the slopes of Mann Hill and offers scenic overviews of the Hoosic Valley. Pownall Center is on the divide between the drainage nets of the Hoosic and Waloomsac Rivers. Pownal Center also lies at the southern tip of the Vermont Valley, in which the bedrock geology controls the topography. The Vermont Valley narrowly separates the Green Mountains on the east, with their core of Precambrian gneisses, from the Taconic Mountains on the west, with their caprock of slates and phyllites. The valley is carved into north-south-trending bands of formations consisting mostly of marbles that have made the Vermont Valley the principal marble producing center of the northeast. The Vermont Valley also served as a convenient channel for glacial advance which left it broad, steepsided, and replete with glacial drift of all kinds, including many economically valuable sand and gravel deposits.

Brecciated carbonate in sharp contact with folded mica schist on US 7, south of Bennington.

Between Pownal Center and Bennington (6 miles), the road skirts the Taconic klippe at the base of the steep east wall of Carpenter Hill. Several cuts expose the marbles of the valley floor that belong to the Ordovician Bascom, and Cambrian Clarendon Springs and Winooski formations. The hummocks of the valley bottom are mostly deposits that look like mounded moraine but are really deltaic sediments deposited in standing water next to the ice. These kames abound between Pownal Center and Pittsford (69 miles) and dozens of new and old borrow pits attest to exploitation of their clean, clay-free materials for use in construction and road-building.

Bennington is well-known for the Battle of Bennington, one of the decisive battles of the American Revolution that marked the beginning of the end for the British campaign. During General Burgoyne's push down the Champlain and Hudson valleys he sent a detachment from Fort Edwards to Bennington to take the town and obtain much needed supplies and horses. Meanwhile, Brigadier General John Stark, in his newly-established command of the 2000-strong Bennington force, got word of the impending assault, and decided to head them off. The actual battle, near Walloomsac Heights in New York, lasted only two hours, but in Stark's words, it was like "one long clap of thunder." As General Stark marched the hundreds of captured and wounded enemy troops back toward Bennington, he was overtaken and surprised by British reinforcements and forced to give ground. Suddenly Colonel Seth Warner arrived from Manchester with his band of Green Mountain Boys; and the tables were turned. The Bennington Battle Monument, completed in 1891, is 306 feet high,

Thrust-faulted bed of dolostone north of Bennington. Road sign indicates general direction of movement of rocks above the thrust. Rocks below overlapping dolostone slab are highly sheared and crushed. Near Maple Hill thrust fault.

the tallest structure in Vermont. It is made of an attractive blue-gray dolostone called Sandy Hill dolomite from Hudson Falls, New York.

Bennington lies in the widest part of the Vermont Valley on the Walloomsac River in a lovely setting with open views of the mountains. The Waloomsac River heads high up in the Precambrian rocks of Glastenbury Mountain northeast of the village, crosses the valley, then cuts a broad swath through the Taconics and joins the Hoosic River at North Hoosic; the Hoosic River continues to join the Hudson at Stillwater, New York.

An excellent roadcut on the northbound lane of US 7 about 1.5 miles north of Bennington serves as a small-scale example of the thrust faulting of western Vermont. The cut bares a faulted 5-foot thick bed of dolostone, probably an interbed in the Cambrian Cheshire quartzite, with the southeastern segment thrust a short dis-

Cheshire quartzite on US 7, north of Bennington.

tance up and over the northwestern segment and with a lot of crushed rock in the fault zone between. This is close to the large Maple Hill thrust, if not the actual fault, which has the same sense of displacement. The overthrust fault slice is an eloquent demonstration of how the continental crust compressed laterally and thickened vertically during building of the ancestral Taconic Mountains. Huge thrust faults riddle western Vermont. They trend from north to south and pile up westward like so many folded bridge chairs stacked against a wall.

Between Bennington and Arlington (exit 3—13 miles), there are a few roadcuts in buff to light gray dolomitic marble of either the Dunham or Winooski formations, both of early Cambrian age. They are close to the Maple Hill thrust and the beds are folded.

Just south of Shaftsbury, the valley is partly blocked by the Hale Mountain moraine that marks a temporary stand of the ice margin during deglaciation. After the ice receded from this position, Flat Top Lake formed north of the moraine. It eventually extended to Arlington where another moraine piled up. Of the two, the Hale Mountain moraine is the smaller, but the most obvious with its hummocky kettled surface and poor drainage. Flat Top Lake lasted for only a short time because the Hale Mountain moraine had an ice-core and the imponded waters spilled out when the ice melted.

Arlington is on the Batten Kill which flows southward in the Vermont Valley to here, then westward through the Taconics and finally joins the Hudson River at Schuylerville, New York. The river is west of the highway between Arlington and Manchester Center (10 miles).

The high mountain west of Manchester Center with wind

Colorfully banded early Cambrian Monkton formation, 1 mile south of Wallingford.

generator towers on it is 3816-foot Mount Equinox. A 5½ mile toll road goes to the summit from Vermont 7A near Sunderland. The views of the Vermont Valley, the Green Mountains, and the Taconics from the top are excellent. From US 7, you can see a large glacial cirque called Skinner Hollow on the east side of the mountain. Hang glider enthusiasts take off from the mountaintop and land in a pasture between routes 7A and 7.

The valley between Manchester Depot and Wallingford (22 miles) is narrow and glacially gouged. The large Manchester Center moraine is around Barnumville. The long, flat stretch of highway north of Barnumville passes many small kettle lakes. Farther north are streamlined drumlin hills made of glacial till and glacially-smoothed bedrock knobs.

Near Danby, several large roadcuts expose the same marble bands as those south of Manchester—the Clarendon Springs, Winooski, and Dunham formations—and there is a large quarry in marble, much of it pure white. A few cuts expose dolomitic marbles and interbedded reddish quartzite of the Cambrian Monkton formation. One mile south of Wallingford is a spectacular, colorfully banded, dark to light gray and brownish to buff cut in the Monkton formation with uniform thin beds that dip about 45 degrees eastward. The rocks are in the Pine Hill thrust slice, which is overthrust by another large fault slice immediately east of the highway. The White Rocks Recreation Area two miles east of Wallingford, by Vermont 140, is in the Cambrian Cheshire quartzite atop the eastern fault.

Between Wallingford and Rutland (10 miles), the Vermont Valley opens wide, and the Taconic and Green mountains are in full and glorious view from the divided highway. Infrequent cuts reveal more of the same rock units exposed farther south. This section of highway crosses the Mendon moraine, which drapes the east half of the valley up to a maximum altitude of about 1500 feet on the Green Mountain slopes. The higher deposits include many kames and most of the sand and gravel pits.

The conspicuous water gap visible east of the highway near Clarendon (5½ miles south of Rutland) is the Clarendon gorge that Mill River carved through the Cambrian Dalton formation and Cheshire quartzite. The Dalton formation lies on the Precambrian rocks of the Green Mountain core at the other end of the gorge. The contact is a profound unconformity representing an erosional gap in the geologic record of about 500 million years, between the time of formation of the core rocks and the deposition of the sediments of the Dalton formation on top of them. The gorge is accessible by Vermont 103, which branches from US 7 two miles north of Clarendon, at Pierces Corner.

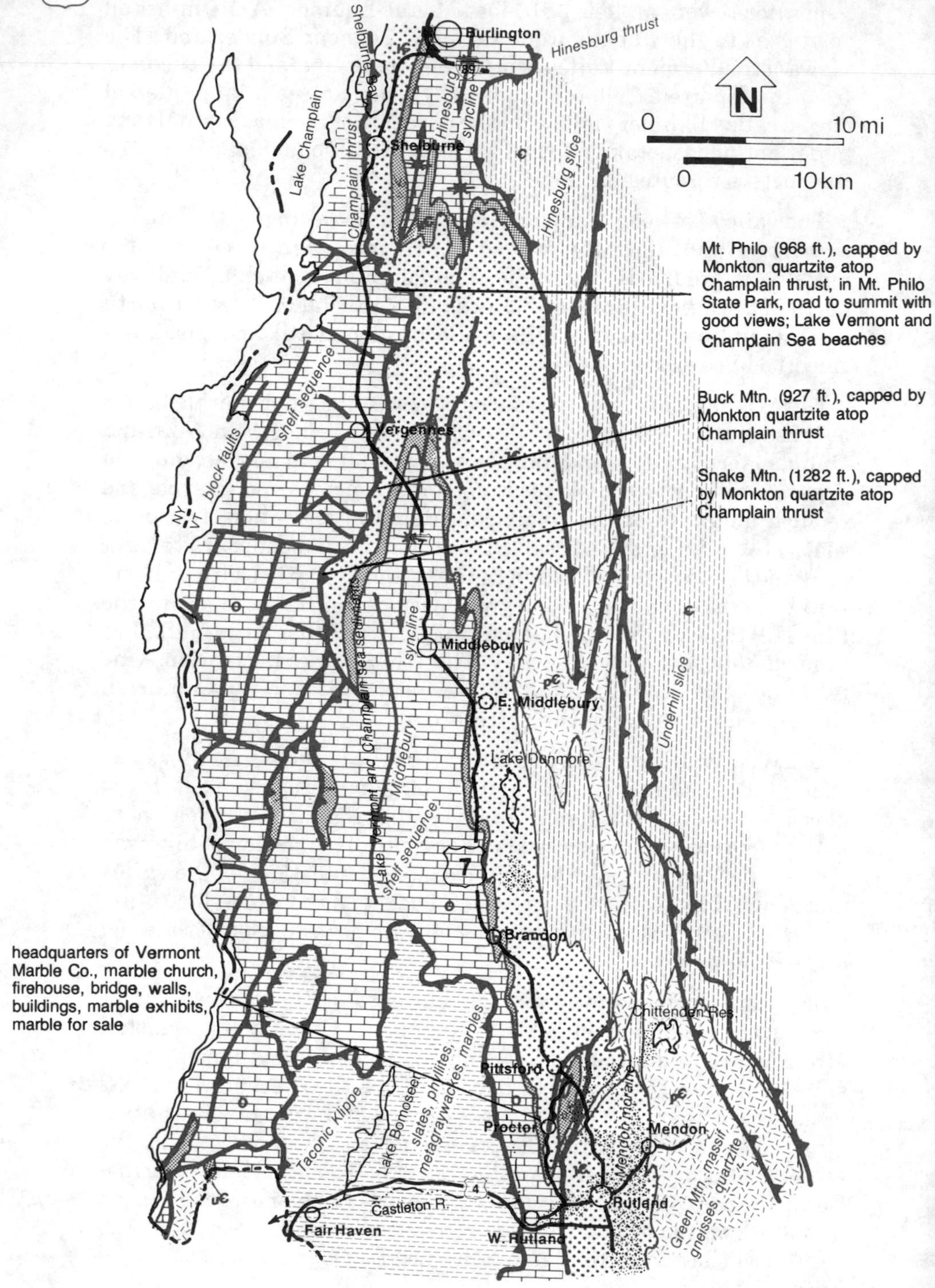

US 7
Rutland — Burlington

US 7: Rutland—Burlington

67 mi. / 107 km.

Rutland, Vermont, second largest city after Burlington, was formerly known as Marble City. The marble industry got its real start here when the stone was first quarried near Proctor in 1836. Proctor later became a separate township that retained the title of Marble City. Near the turn of the century, Proctor was one of the major marble centers of the world when the Vermont Marble Company had quarries throughout the state, as well as in Colorado, Tennessee, and Alaska. The city is by Otter Creek, near the northern end of the broad section of the Vermont Valley that extends south ten miles to Wallingford. It is a scenic area with the Green Mountains to the east, the Taconic Mountains to the west. This broad section of the valley owes its width to broad marble bands in the valley floor that erode more easily than do the siliceous rocks of the valley sides. The first line of hills east, north, and west of the city are all carved from a "horseshoe" band of Cheshire quartzite that opens to the south.

Rutland lies at the edge of the Mendon moraine, the largest and most continuous moraine belt in western Vermont. The deposits are really in two parts with till moraine on the valley floor and kame-moraine full of water-deposited material on the eastern side. The moraines drape over the foot of the mountains up to an elevation of 1500 feet. Many sand and gravel pits operate in the kame-moraine part. The landscape around Rutland is conspicuously morainal, full of mounds and hollows that become more subdued south of the city.

Between Rutland and Pittsford (9 miles), US 7 crosses the Pine Hill thrust fault and another branching fault. Some roadcuts expose quartzite, a hard, brittle rock, that appears to be shattered by stresses associated with movement along the fault.

The Vermont Valley ends at Brandon, 9 miles north of Pittsford, opposite the northern tip of the Taconic klippe and blends into the broad lowland of the Champlain basin. The relatively flat topography northward to and beyond Burlington results partly from deep glacial scour and partly from sediments deposited in Lake Vermont and the Champlain Sea. Some of the higher rock knolls that jut from the surface were islands in Lake Vermont. Well-preserved shoreline features mark their slopes. At maximum depth, Lake Vermont lapped against the steep eastern scarp of the Green Mountains east of US 7 and extended a long finger to South Wallingford in the Vermont Valley. The dry beaches and other shoreline features evident there and on the islands now increase in altitude toward the north because postglacial uplift increases in that direction. Actually the surface of the Champlain basin is not as flat as it was originally because the sediments laid down in Lake Vermont and later in the much smaller and shallower marine embayment called the Champlain Sea have been extensively dissected by streams. The streams cut downward as the land under them rose in postglacial rebound.

Brandon is famous for the Brandon kaolinite deposit about 2 miles northeast of the village. Kaolinite is an earthy white clay, though often discolored by impurities, formed principally by weathering of feldspar. The alteration of feldspar in Precambrian gneisses of the Green Mountains 20-30 million years ago apparently supplied the clay and streams transported it to this site. Vermont then had a warm climate similar to that of the southeastern United States today and this area was swampy. The clay locally buried much swamp debris now evident in well-preserved fossil nuts, fruits, leaves, and wood which, where abundant, form lignite beds. Also northeast of Brandon is the so-called Stucco Pink Quarry which produced pink marble from the Cambrian Dunham formation.

Between Brandon and Middlebury (18 miles), the highway follows the east limb of the Middlebury syncline. Few roadcuts expose the underlying Ordovician and Cambrian marbles. The marshy strip of land containing Otter Creek, 4-5 miles west of the highway, is in the trough of the fold. The prominent, continuous ridge east of the road called the Hogback range, is carved from steeply dipping beds of early Cambrian Cheshire quartzite. Basically, these beds are upfolded over the Green Mountain core to the east and downfolded into the Middlebury syncline to the west. The ridge is the erosional remnant of the steepest part of the limb between the two folds. It stands high because it is composed of harder and more resistant rocks than the marble bedrock west of it.

Despite the historical prominence of Rutland and Proctor as the focus of the marble industry, Middlebury became the site of the first extensive quarry in 1803. The city really has a marble foundation—several bands of Ordovician marbles pass through it close to the axis of the Middlebury syncline.

Between Middlebury and Vergennes (12 miles), there are open views to the west that include two high hills. Snake Mountain (1282 feet) six miles northwest of Middlebury and Buck Mountain (927 feet), three miles west of New Haven Junction. Hard Cambrian Monkton quartzite overlying the Champlain thrust fault, which emerges along the west side of the Middlebury syncline, holds up both. Rocks beneath the fault are relatively weak Ordovician marbles that were tightly folded as the fault moved. The fault has been traced northward to the Canadian border, southward nearly to the end of Lake Champlain. The highway crosses the fault near Vergennes along the north slope of Buck Mountain. East of Buck Mountain and highly visible from the road is an impressive narrow slot in the Hogback range where New Haven River issues from the mountains.

Many little bedrock hills near Vergennes have been shaped, ground smooth, and scratched by overriding ice. Some have the characteristic sheepback form of gentle north slope and steep south slope. Such hills are common throughout the Champlain basin but seem to be more conspicuous here. Deep sediment deposits may bury many more.

Between Vergennes and North Ferrisburg (6 miles), roadcuts expose the Chazy limestone that lies just west of the Champlain thrust. The gray-to-buff rock is intensely deformed and laced with white calcite veins. Parts of the Chazy limestone are highly fossiliferous; but near the fault and east of it, most of the fossils have been destroyed by deformation and metamorphism. The fossil for which this unit is most famous is *maclurites magnus*, a flat-coiled snail that grew to several inches in diameter.

The lone, almost conical hill just west of the highway and a mile north of North Ferrisburg is Mount Philo, the centerpiece of Mount Philo State Park. This was also an island in Lake Vermont. A short, steep road leads to the summit of this little mountain, to one of the best panoramas of the Champlain Valley. The view encompasses the flat lowland with its rock islands, a great expanse of the lake, the Adirondack Mountains on the other side of the lake, and the Green Mountains in the opposite direction. There is much more.

The mountain is supported by a caprock of Monkton quartzite above the Champlain thrust fault, resting directly on the erosionally weak Ordovician Stony Point shale below the fault. Several roadcuts on US 7 north and south of the park entry road expose highly deformed, shingled sections of the Stony Point shale. The Mount Philo

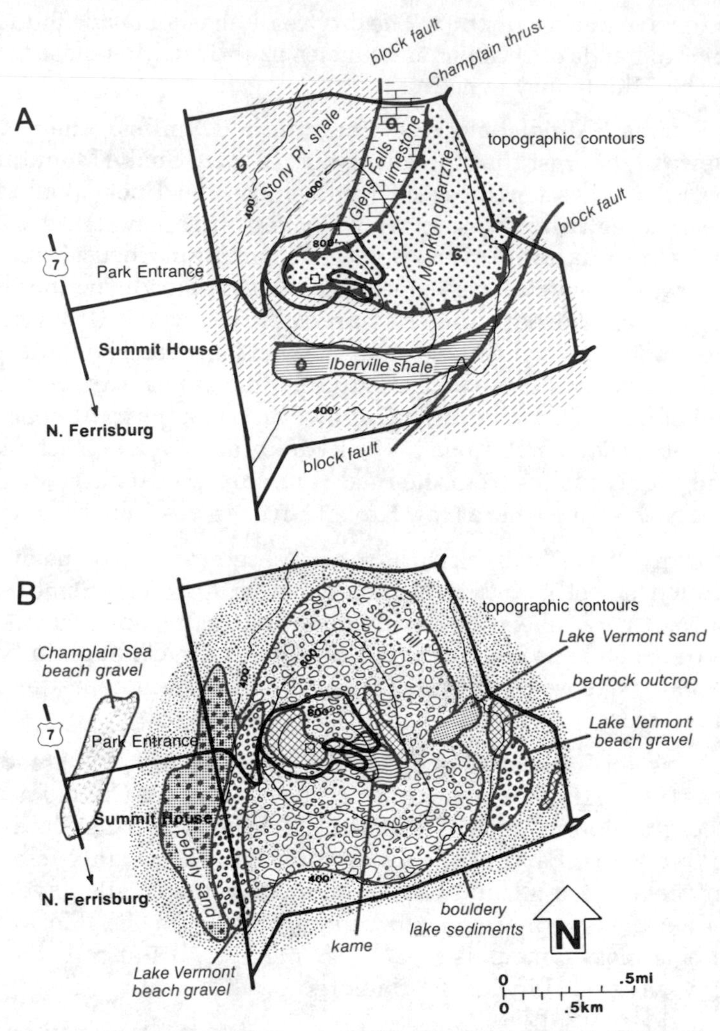

Mount Philo State Park (A) Bedrock geology and (B) surficial geology.
—Adapted from H. W. Dodge.

summit lookout is on the maroon quartzite of the Monkton scarp. Stony Point shales are exposed below the scarp near the end of the one-way summit auto loop. The steep slope above these exposures is strewn with angular blocks of quartzite that have fallen from the cliff, which repeatedly collapses as the weaker supporting rock is undercut. The same mechanism maintains the free-fall of Niagara Falls.

The slopes of Mount Philo are mostly covered with glacial drift, including a lot of till, Lake Vermont sediments, a kame, and beach gravels. Lake Vermont beach deposits are exposed by the low switch-

View south from Mount Philo over Champlain lowlands. The three bumps on the horizon are fault scarps capped by Monkton quartzite as on Mount Philo itself. The middle bump is Buck Mountain and the farthest one is Snake Mountain near Chimney Point. These mark the leading edge of the eroded Champlain thrust slice.

back on the two-way section of the summit road at an elevation more than 500 feet above Lake Champlain. Marine beach gravels marking the late Champlain Sea shore exist at lower elevation by the US 7 intersection directly west of Mount Philo. Seawater apparently did not reach high enough to wrap around the hill and make an island of it, but it came close.

Between Mount Philo and Shelburne (7 miles), the road crosses the Champlain thrust fault at least twice. Some roadcuts display Monkton quartzite of the overthrust sheet, others reveal the underlying Ordovician shales. All of the rocks near the fault are intensely deformed. North of Shelburne, the fault crosses under Shelburne Bay, which is visible from the road in the last few miles to Burlington.

Marble block administration building of the Vermont Marble Co. at Proctor with sawn marble slabs in foreground.

A Side Trip to Proctor, The "Marble Center" of Vermont

The pretty little village of Proctor nestles in narrow Otter Creek Valley 6 road miles northwest of Rutland. It is easily accessible by way of Vermont 3, the Marble Valley Highway, which goes north from US 4 two miles west of Rutland. Proctor is the "Marble Center" of Vermont, built around the headquarters and finishing plants of the state's principal marble industry.

Quarrying of marble in the vicinity began in 1836 when what is now Proctor was part of Rutland. By 1870, there were numerous quarries in the region, each owned by a different, small company. The largest quarries were in West Rutland, Florence, Brandon, and Sutherland Falls (Proctor). In that same year, Colonel Redfield Proctor bought the Sutherland Falls Marble Company and began to absorb most of the other companies, welding them into the Vermont Marble Company. The company continued to expand, becoming one of the giants of the industry in its heyday during the late 19th and first half of the 20th centuries. In 1886, Redfield Proctor influenced the legislature to partition the town of Proctor from Rutland—he and his family then owned, or controlled, about 97% of the property in Proctor.

Today, widespread use of cheaper synthetic products has greatly diminished the marble industry. Vermont Marble Company survives as a small cut-stone division of a large company that markets crushed marble from some of the old quarries for use as paper coatings, extenders, food additives, fillers, and the like. The village remains a virtual monument to the illustrious past. The company headquarters are there as are their exhibits depicting the geology of Vermont marbles and displays of marble from all over the world. There is also a gift shop and a marble market where you can buy marble and verde antique (serpentine) from Rochester, Vermont, at factory prices. Marble is everywhere. The beautiful old buildings in the company complex, built of marble blocks used as concrete cinder blocks are today, include the administrative office building, the old machine shop, and part of the present mill. The polished marble facade of the nearby firehouse is a memorial to Mortimer R. Proctor. A graceful arch bridge of marble spans Otter Creek. The most strikingly beautiful marble structure is St. Dominic Catholic Church. There are marble gravestones, homes, walls, steps, even sidewalks. Most of the stone curbings are granite from Barre, Vermont—a nice contrast.

The Vermont Marble Company still operates only three quarries: the underground Imperial Quarry at Danby, the open pit verde antique quarry at Rochester, and the open pit quarry in fossiliferous Crown Point limestone—really marble—on Isle La Motte. Only the Imperial Quarry operates all year. The Rochester quarry is only active from April to November, the Isle La Motte quarry only when the company has special orders for its stone.

The Imperial Quarry is in Dorset Mountain west of US 7 between Manchester and Danby. The company claims that it is the largest underground marble quarry in the world. The marble now being taken is in thick, nearly horizontal layers overlain by a 9-foot-thick bed of worthless rock. Quarries first strip the worthless overburden. Then, working from the top of the marble, the quarrymen drill "channels" of vertical one-inch holes spaced three inches apart. Next, they drill a similar line of horizontal holes into the quarry face at the

Marble wall in Proctor.

Imperial marble quarry in Dorset Mountain.
—Courtesy Vermont Marble Co.

base of the block. Finally, they gently split the block off by driving wedges into the holes. Explosives would shatter the fragile marble. The blocks normally measure about 5 feet square and 8 feet long and weigh about 18 tons. Larger blocks can and have been taken for special orders: among them one 93-ton monster used for a covered wagon sculpture in Salem, Oregon, and an 83-ton block used for a sculpture at the Veterans Memorial Building in Detroit.

A huge forklift truck capable of lifting 40 tons picks up the separated block and takes it to a storage area near the mine entrance where a flatbed truck will pick it up and carry it to Proctor. At the mill, the blocks are cut, polished, engraved, sculptured, and otherwise processed.

The Imperial Quarry now covers more than 20 acres with vast open spaces and high ceilings deep within the mountain. Evenly spaced piers of intact marble, measuring about 30 x 30 feet and spaced about 40 feet apart, support the roof.

E. L. Smith quarry in Barre granite showing system of derricks.

A Side Trip to the Barre Granite Quarries

The region a few miles southeast of Montpelier has long been famous for its quarries in the so-called Barre granite. Of the many granite bodies in New England, the Barre is one of the best known because it yields monument stone of exceptionally high quality. It is one of the smaller New Hampshire series plutons, deep-seated igneous intrusions. The much larger Knox Mountain pluton begins a few miles east of Barre and extends northward for about 20 miles. Many more occur in the northeastern corner of the state and in New Hampshire. All are late Devonian to early Mississippian in age, and all were apparently emplaced as molten magma during and immediately following the Acadian mountain-building event, which also imprinted the intruded country rocks with deformation and metamorphism.

Barre granite is an attractive, light gray rock in which slight differences in shading are controlled by the amount of dark biotite mica it contains. One of its most distinctive features is its unusually uniform, medium-grained texture. Stone removed from almost any part of the quarry can be expected to be almost identical to that of any other part. These characteristics coupled with the fact that the cut stone takes an extremely fine polish make it ideal for use in monuments.

The Rock of Ages pit is a tremendous hole in the ground measuring about 1200 feet long by 600 feet wide at the top and about 400 feet deep. Quarrying involves the separation of oversize, rectangular blocks of granite. The blocks are freed first by making vertical "channel cuts." In the past this was accomplished by drilling a line of two-inch holes and then drilling out the one-inch section between the holes. Now most is done by a process called "jet-channeling," by which a high velocity jet of fuel oil burning in pure oxygen slices through the granite by heating it to 4000 degrees Fahrenheit and causing it to "flake." Though more expensive, the process is more efficient, and it makes a smoother cut. In the early days of quarrying, much stone was removed by blasting which does far more damage. Many of the oldest cuts in the quarry face show blasting scars. Horizontal cuts, or "lifts," are made by drilling a line of holes a foot apart and then blasting with black powder. The blast is so mild that it does not shatter the granite, but merely splits it along the line of the holes.

Once freed, the blocks are further reduced to a more manageable size by a third splitting technique. Using pneumatic drills, quarrymen drill a line of small diameter, shallow holes. They then place two shims and a wedge in each hole and gradually and evenly drive the wedges with a striking hammer to split the rock apart. In all these operations the quarrymen try to follow the natural structures in the rock to facilitate splitting and give the highest quality stone.

Smaller blocks are hoisted from the pit by one of the several derricks mounted around the quarry. Each consists of a mast (about 115 feet high) and a boom made of enormous Douglas fir timbers from Oregon. The mast is held upright by a system of steel guy wires anchored either to the bedrock, or to "deadmen" consisting of large blocks of granite buried in the ground. The derricks have a capacity of 40 to 50 tons, but the blocks removed from the quarry average only 20 tons. The hoists also lower the quarrymen to their work stations and bring them back up. Each time a new section of the pit is started where there is no derrick, one has to be dismantled, moved, and set up again—a formidable engineering task.

Blocks are sawn into slabs of varying sizes at the sawplant near the quarry. Cutting is mainly done by a wire saw consisting of a long strand of twisted wire. The wire, which has reversed twists every 50 feet, travels over a system of pulleys and through a slurry of silicon carbide abrasive before contacting the stone. The abrasive lodges in the wire grooves; and it, rather than the wire itself, effects the cutting.

Sawn slabs go to the Craftsman Center, where they are further shaped, polished, and engraved to order. Only about 15% of the quarried stone is marketed. The rest, discarded for imperfections, makes up the huge waste piles, or "grout," you see all around the many pits in this region.

Quarrying of Barre granite began soon after the War of 1812. The state capital building at Montpelier was made of this stone in the 1830s. The real beginning of the granite industry in this region, however, came after the Civil War with the coming of the railroads, which greatly facilitated transporation of the heavy stone.

Stony bed of Wells River, west of Groton, Vermont. Incredibly, even largest stones may be moved in spring floods.

US 302: Barre, Vt.—Maine Border

109 mi. / 174 km.

see map page 206

From the air, the city of Barre looks like no more than a cluster of buildings and long sheds by the railroad, dominated by the shattered scars of nearby Millstone Hill where granite has been quarried for 100 years. The industry began after the War of 1812 with quarries on Cobble Hill that produced stone for millstones, doorsteps, posts, and window lintels. A finer grade was later found on Millstone Hill where almost all of the quarrying has been done since. Despite all that, Barre is a city with only a few granite buildings. Instead, many homes are Georgian brick, and some public buildings are brick with granite trim. The Robert Burns memorial statue, which faces the central plaza from the high school grounds, is a beautiful granite carving, and there are others in the cemeteries.

US 302, between Barre and Wells River, at the New Hampshire border (31 miles), cuts across the major north-south-trending deformational features of eastern Vermont. This region lacks the big, east-dipping thrust faults that distinguish western Vermont. Instead, it contains giant synclines and anticlines that stretch north to south nearly the length of the state. Just east of Barre, the route cuts across the northern tip of the Barre granite pluton of the Devonian New Hampshire series, on the axis of the Townshend-Brownington syncline. Near the village of Orange, about seven miles east of Barre, the road touches the exposed southern tip of the much larger Knox Mountain pluton at the axis of the Strafford-Willoughby arch. Near Wells River, you cross the axis of the Brattleboro syncline. The contorted rocks visible in the scanty roadcuts are mostly thin-leaved, dark-colored schists, marbles, and quartzites of Devonian age belonging to the Waits River and Gile Mountain formations. Since their age is post-Taconian, their folding and metamorphism is obviously Acadian. The grade of metamorphism increases from either side to a maximum near the Strafford-Willoughby arch. Rocks there contain metamorphic index minerals kyanite and staurolite. The higher grade is due to contact metamorphism next to the intrusive bodies emplaced during regional Acadian metamorphism. Where no plutons are exposed in this axial zone, they probably lie at shallow depth. In northeastern Vermont, other granite bodies in the New Hampshire plutonic series have cooked the country rocks even more, resulting in crystallization of the higher temperature index minerals sillimanite and andalusite. At the eastern end of this traverse, you cross about five miles of low-grade greenish-gray slates, phyllites with lustrous surfaces, and quartzites belonging to the Ordovician Albee formation.

The Connecticut River marks the boundary between Vermont and New Hampshire over the full length of the states. The original position of the river appears to have been influenced by the west-dipping Ammonoosuc thrust fault, which nearly parallels it for about 25 miles, between Lancaster, New Hampshire, and Windsor, Vermont. North of Lancaster, the river rises to its headwaters at the Connecticut Lakes of northern New Hampshire. South of the states, in Massachusetts and Connecticut, the river winds its tortuous way over the broad floor of the Connecticut Valley basin, one of a string of large fault-block depressions near the eastern margin of the Appalachian chain. All are products of crustal rifting that began about 200 million years ago, in early Mesozoic time, and ultimately led to the opening of the present Atlantic Ocean basin.

The Connecticut Valley was extensively modified by Wisconsin glaciation. Erosion during ice advance was followed by deposition of till, outwash, and glacial Lake Hitchcock sediments during ice retreat. Downcutting of the river and its tributaries followed as the land rebounded after glacial retreat. The net result is a valley distin-

Bretton Woods resort with Mount Washington behind.

guished by Lake Hitchcock shoreline features and terraced sides with exposed sections of kames, deltas, outwash and varved clays. In the vicinity of Woodsville are numerous varved clay exposures, one of which records 1600 years of accumulation!

Between Woodsville and Littleton (22 miles), the road follows the Ammonoosuc River along a rather flat valley. Scanty roadcuts expose thinly-leaved greenschists of the Ammonoosuc volcanic pile, part of the Bronson Hill island arc complex. The west-dipping Ammonoosuc fault lies just to the west. A few of the roadcuts are glacially smoothed and striated at their tops. A narrow section of the valley south of Lisbon displays hummocky morainal topography with scattered erratic boulders.

Between Littleton and Twin Mountain (13 miles), you cross the southwestern end of the largest of the Ordovician Oliverian plutons, the Jefferson dome, on the axis of the Bronson Hill anticline. One good roadcut, about two miles east of Littleton, exposes the coarse-grained gray gneiss of the pluton. Just east of Littleton, the road climbs steeply out of the Ammonoosuc Valley. Much of the way to Twin Mountain is open with good views of the mountains all around. Especially prominent on the eastern skyline are the high, barren summits of the Presidential Range. Between Twin Mountain and Bretton Woods (5 miles), the road crosses granite of both the White Mountain and the New Hampshire plutonic series. Crawford House, three miles south of Bretton Woods, stands on the divide between the Ammonoosuc and Saco River systems, between tiny Amonoosuc and Saco lakes at their headwaters. From here, US 302 follows the Saco River to the Maine border (35 miles). Most of the bedrock is pinkish, coarse-grained Conway granite and other rocks of the White Mountain magma series. Roadcuts are scarce, but glacial features abound.

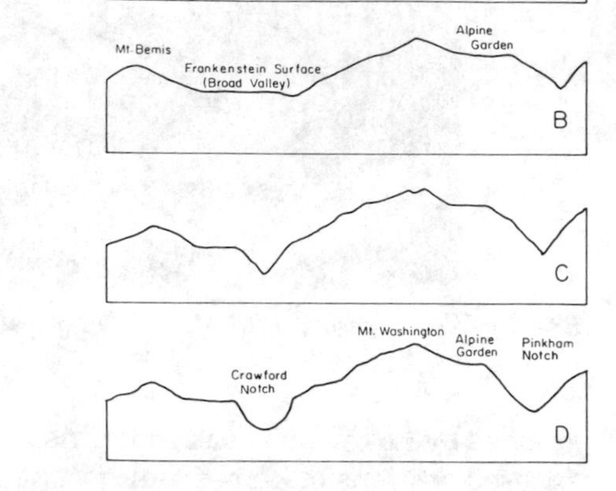

Evolution of Crawford and Pinkham notches and other erosional features in Mount Washington region: (A) carving of Presidential upland surface near sea level; (B) after uplift of several thousand feet and stream erosion; Alpine Garden is remnant of upland; (C) renewed uplift of 1000-1500 feet caused further stream incision; (D) notches are glacially reshaped during Ice Age.
—Adapted from M. P. Billings.

The north entrance to Crawford Notch, called the "Gate of the Notch," lies just south of Saco Lake. The section referred to as the notch is only about three miles long, but it is a superb example of a glacially-gouged valley. Another equally spectacular but smaller one is Zealand Notch, four miles farther west. Unlike the similar valleys of the upper slopes of the Presidential Range with bowl-shaped rock amphitheaters at their heads, these were not hollowed out by isolated, alpine glaciers. Instead, ice tongues at the base of an overriding ice sheet gouged out the existing stream valleys. Many other similar valleys more or less aligned with ice flow exist in Vermont and New Hampshire.

The funnel-like Gate of the Notch may have contributed to the perfection of its form. Ice must have been crowded into this narrow opening like toothpaste being squeezed out of a tube, even before it overtopped the high ridges on either side. Mount Willard, a 1500-foot high knob of Conway granite that stands right in the way at the northern entrance, deflects the valley eastward, and complicates this picture. The ice tongue had to go around a tight double bend to enter the main valley. Overriding Wisconsin ice eventually carved Mount Willard into a giant "sheepback" with a gentle north slope and a steep cliff facing the main valley to the south. The cliff results from glacial plucking of rock from the unsupported downstream side of the mountain. Mountainsides flanking the gate were also smoothed by the ice onslaught.

Best views of Mount Willard and the valley are from the Willey House Camps, by a small pond about midway through the notch.

From here you can see the steep, talus-draped upper slopes and cliffs of Mount Webster on the east side and the equally steep west side. These slopes are prone to slide. The most famous landslide occurred on August 28, 1826, at the site of the Willey House, 1½ miles south of the Willey House Camps at the southern end of the main valley. The slide debris is still visible there nearly blocking the valley. Saturation of the soil by a violent storm, which also raised the Saco River to 24 feet, triggered the slide. This disaster killed nine members of the Willey family while sparing their house as the slide split just above it.

Between Willey House and Bartlett (10 miles), glacial reshaping of the Saco Valley by the Crawford Notch ice tongue is much less apparent. This section of the valley is broad and the walls severely dissected by small streams tributary to the Saco River. Three miles west of Bartlett, the valley makes a right angle bend to the east athwart the Wisconsin ice flow. Here, ice sculpturing differs from that of the notch. The valley floor is broad and flat; the north side is steep and cliffy as a result of glacial plucking; and the southern slopes are relatively gentle and smooth. The Saco River meanders widely over the flat, sediment-filled valley floor. Conditions are similar between Bartlett and Glen (6 miles).

The road turns sharply south-southeastward at Glen. Between Glen and Redstone (9 miles), it follows the heavily-developed, touristy Mount Washington valley with the mountain visible to the north. From Redstone to the Maine border (7 miles), it goes eastward again. This whole section of the Saco Valley is extremely broad; and the river meanders even more widely than before. Much of it is open; the mountain scenery is superb, punctuated by domical granite summits that are a trademark of New Hampshire. The flat-floored New Hampshire section of the Saco Valley, at least from Glen to the Maine border, is filled with glacial outwash and lake sediments. Numerous borings have revealed the great depth of sediment fill; one in Conway, for example, found 95 feet of sand, with clay layers in the upper 25 feet. Apparently the valley fill was originally much deeper since postglacial downcutting and meandering of the Saco River has removed much of it. The striking river terraces on the valley sides near North Conway result from this action.

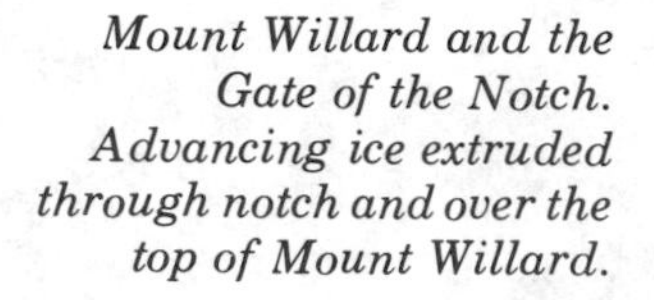

Mount Willard and the Gate of the Notch. Advancing ice extruded through notch and over the top of Mount Willard.

VT 9
Brattleboro — New York Border

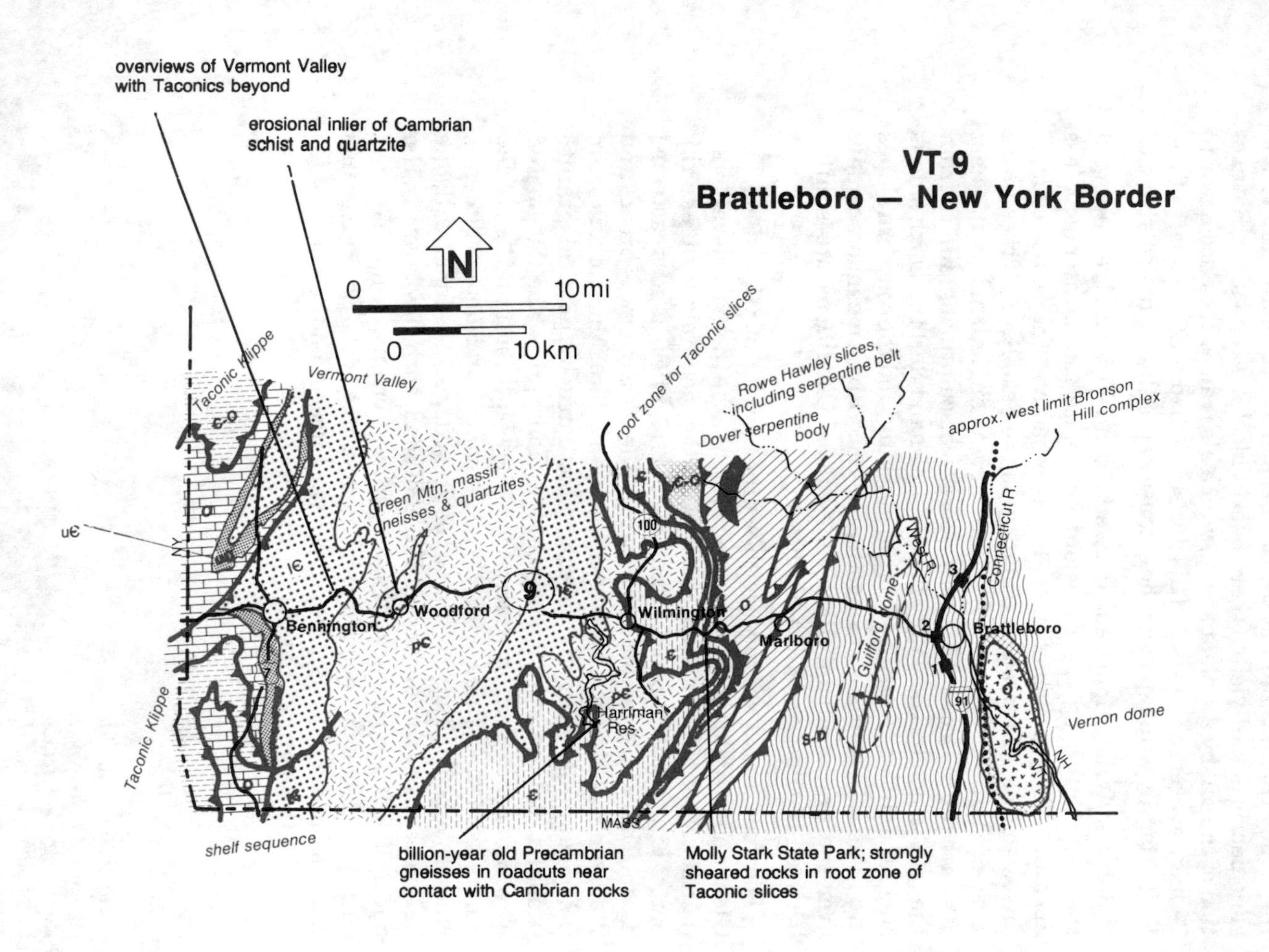

Vermont 9: Brattleboro, Vt.— New York Border

39 mi. / 62 km.

The Molly Stark Trail

This east-to-west route crosses the narrowest section of southern Vermont where the north- to south-trending folds and faults of northeastern Vermont are most tightly crammed together. The route offers a view of major rock and deformational features that characterize much of the state in a short distance. In map view the whole resembles a curtain tied at the bottom on the left side of a window.

Between Brattleboro in the Connecticut Valley and Marlboro (7 miles), you follow Whetstone Brook. This section is rather woodsy, but a few roadcuts expose phyllites, greenschists and schists with knots of magnetite or garnet, and micaceous quartzite of the Devonian Waits River formation, all with steeply dipping layers reflecting the tight folding. You cross into the highly sheared Ordovician rocks of the serpentine belt just east of Marlboro. The belt contains numerous slivers of dark rocks—dunite, peridotite, or gabbro—that were ripped from the ocean crust in the early stages of Taconian mountain-building, dragged upward along faults, and altered in varying degrees to serpentine and talc. None are readily apparent along Vermont 9, but the largest of the entire belt, called the Dover body, is at East Dover, a few miles north and a little west of Marlboro. Hundreds more are farther north, with the largest cluster located around Troy near the Canadian border. Many are good mineral collecting sites, with exotic minerals not found elsewhere in Vermont or New Hampshire, or in any other places.

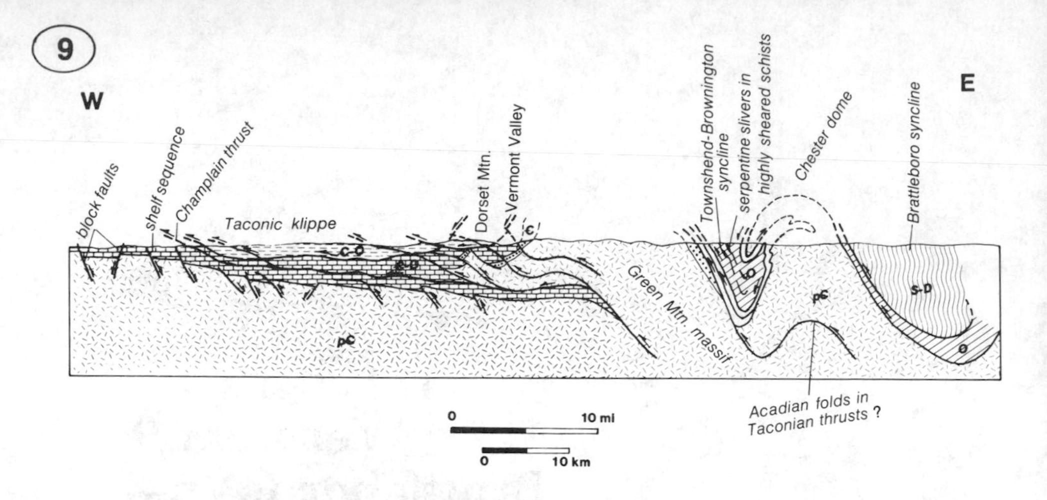

Schematic east-west cross section south of Rutland, showing major elements discussed in the text. —Adapted from R. Stanley and N. Ratcliffe.

Practically all of the rocks in the serpentine belt are highly sheared and, in this part at least, steeply dipping. Between Marlboro and Wilmington (8 miles), the road crosses closely-spaced major thrust faults that follow the eastern border of the Precambrian gneissic core of the Green Mountains. The thrust slices were stacked westward and upfolded over the core, and then eroded from the crest of the anticline, exposing the gneisses. This is the root zone for the Taconic thrust fault slices, and the Taconic klippe is an enormous erosional remnant of their leading edges. Nearly all the roadcuts display intense shearing and tight minor folding. The rocks are various schists and dark amphibolite formed through metamorphism of basalt of Cambrian to Ordovician age.

Wilmington is a ski town near the north end of Harriman Reservoir, a long narrow reservoir in a section of the Deerfield River dammed near Whitingham. Roadcuts by the north end of the lake west of Wilmington expose pale-gray, micaceous Precambrian gneiss with lenses, layers, and streaks of white feldspar. The rocks are of Grenville age, like those of the Berkshire Hills of Massachusetts and the Adirondacks and Hudson highlands of New York; they are at least 500 million years older than the oldest Cambrian schists seen along the road east of Wilmington. Numerous other roadcuts in Precambrian gneisses are between Wilmington and Bennington (21 miles) within the Green Mountain massif, along with a few in micaceous quartzites and marble. Woodford (14 miles west of Wilmington) lies within an erosional inlier of Cambrian Dalton formation schists and quartzites surrounded by Precambrian rocks.

Between Woodford and Bennington (8 miles), the road descends steeply to the Vermont Valley through a gorge cut by City Stream, which eventually drains to Walloomsac River through Bennington. Near the base of the steep incline, you pass into the Cambrian Cheshire quartzite that supports the high ridge north of the road. In roadcuts, this unit appears pinkish buff, well-bedded, and brittle, as indicated by a shattered appearance. Some sections of the unit contain dolostone interbeds.

Open views of the lovely Bennington section of the Vermont Valley are especially impressive along this eastern approach. The Taconic Mountains form the west wall, and the great dome of the Green Mountain massif the east wall. The broad slash in the Taconics directly west of Bennington is the valley of the Waloomsac River, which flows west to the Hudson River.

The gently rounded verdant Taconic summits north and south of Bennington conceal one of the most dramatic geologic stories of the eastern United States, the story of the Taconic klippe. The klippe is composed of out-of-place rocks that were bodily transported westward for many miles by thrust faults. That was part of the immense compression associated with the closing of the proto-Atlantic Ocean, a compression that resulted first in the Taconian, and millions of years later, the Acadian mountain-building events. At least some of the faults initiated during the Taconian event appear to have been reactivated in the Acadian episode, and thrust slices were also folded in this final compression. As might be expected this resulted in a complex of stacked thrust slices containing highly deformed and metamorphosed rocks. Subsequent erosion during more than 300 million years amputated this complex from its eastern roots leaving a mass of strange rock of astounding proportions: 165 miles long from north to south by 10-20 miles wide from east to west by thousands of feet thick resting atop rocks, mostly carbonates like those of the Vermont Valley, to which they bear no genetic relationship.

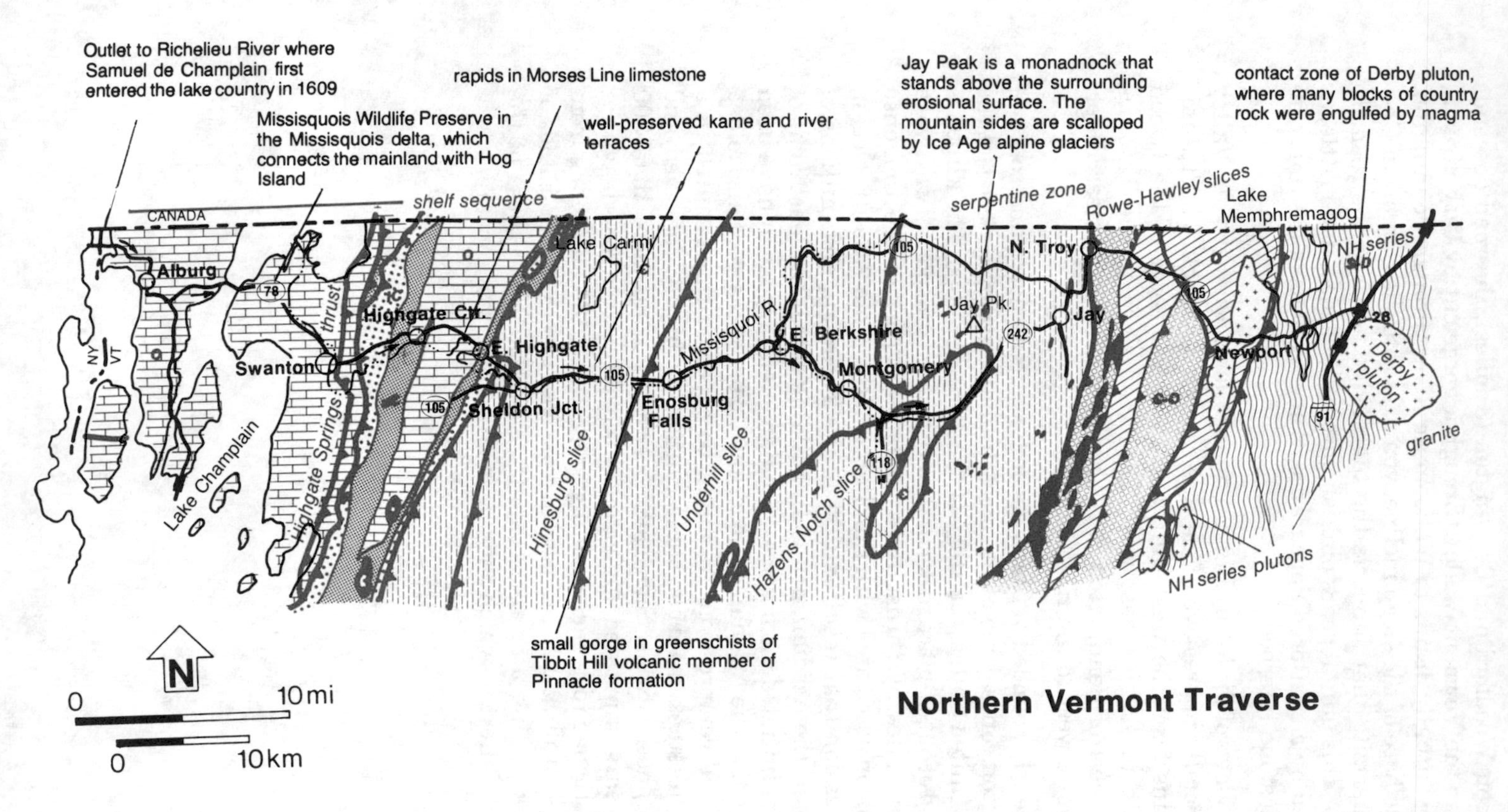

Northern Vermont Traverse

Vermont 78, 105, 118, 242: Northern Vermont Traverse Alburg—Derby Center

74 mi. / 118 km.

The Alburg peninsula projects southward into Lake Champlain from Canada between the Richelieu River and Missisquoi Bay, and farther south between Isle La Motte and North Hero Island. From here, the Richelieu flows almost straight north for 75 miles to Sorel, Quebec, where it joins the St. Lawrence River at a point between Montreal and Quebec City. It was in 1609 that Samuel de Champlain made his way up the Richelieu to the lake he named for himself.

The Alburg landscape is rather featureless, but with good reason. In the first place the Champlain basin is a structural lowland, bordered on the east and west by relatively upthrown blocks of crust. On the Vermont side are the west-directed great thrust slices that override the bedrock of the Champlain floor and account for the high north-south ridges visible from this section of road that culminate in the Green Mountains. Also visible from the road are the Adirondack Mountains on the New York side; these are bordered by numerous, steeply-inclined block faults that dropped the basin side. Secondly, the land was deeply scoured by ice lobes that advanced southward through the Champlain Valley during each of the major glaciations of the Ice Age.

Another reason for the muted landscape here and throughout the Champlain basin is a blanket of lake sediments. In the closing stages of the Wisconsin glaciation as the margin of the Champlain ice lobe receded northward, glacial Lake Vermont filled the basin to levels much higher than that of the modern lake, to shorelines that extend much farther inland. This peninsula and the Champlain islands were

under water. Later, when the ice melted still farther north, Atlantic seawater flowed in through the St. Lawrence and Richelieu valleys to form the Champlain Sea. The result is a deep mantle of clay and silt.

Rapids at East Highgate. Beds dip to the left.

Boulders strewn about some of the Alburg fields are glacial erratics, transported from the north and dumped in place. Most were exhumed from their cover of sediments by wave erosion that accompanied gradual lowering of the Champlain Sea. Subsiding water level was mainly due to rebound of the land surface that, at the height of glaciation, had been deeply depressed by the great weight of the ice. Shoreline features of Lake Vermont and the Champlain Sea are evident throughout the basin; the highest are now several hundred feet above the lake at its northern end, the uplift due to rebound.

Between east Alburg and Swanton, Vermont 78 crosses Hog Island

and then follows the Missisquoi River through Missisquoi Wildlife Refuge. The refuge is in the broad expanse of swampy delta that now bridges the strait from the eastern shore of the lake to the island. The Missisquoi is the northernmost of three large rivers of northern Vermont that cut completely across the Green Mountains from east to west. The others are the Lamoille, next to the south, and the Winooski, still farther south. Each, at an earlier, higher lake level, built a large delta, that is now trenched by the modern stream.

The Champlain thrust fault passes north-south through Swanton with little surface expression. The route between Swanton and East Highgate rises gradually as it crosses the St. Albans syncline, a broad, open downfold of Cambrian and Ordovician beds between the Champlain and Hinesburg thrust faults. A small roadcut at Highgate Center exposes upturned beds of dark Sweetsburg slate on the west limb of the syncline. East Highgate is at the base of the prominent Hinesburg thrust fault scarp held up by resistant Cambrian Cheshire quartzite. Along this route the ridge appears as a moderate rise in the approach from the west; it is higher and far more dramatic farther south at St. Albans. The road here does not climb the scarp, but passes through it along the Missisquoi Valley. There are rapids at East Highgate at times of high water, where corrugated ridges of Morses Line limestone and dolostone beds cross the stream. This unit is of middle Ordovician age, younger than the Cheshire quartzite, yet lies beneath it because of the overthrusting. However, all of the rocks above and below the thrust fault have been shoved westward over the Champlain floor, which is referred to as the "foreland." The rocks below are in the Champlain thrust slice; the folding was contemporaneous with thrusting.

With the exception of the narrow strip of Cheshire quartzite, all of the rocks traversed between East Highgate and Jay (34 miles) belong to the Camels Hump group, chiefly comprised of dark-colored schists and phyllites formed by medium-grade metamorphism of clay muds, dirty sandstones, and volcanic materials. Many of the rocks are green, owing to the presence of the green metamorphic minerals chlorite, chloritoid, actinolite, or hornblende. Carbonate rocks—marbles—are subordinate, indicating a scarcity of limestones in the original sedimentary deposits. Most of the marbles are west of the Hinesburg thrust in the so-called "shelf sequence" formed from carbonate deposits of the ancestral continental shelf.

In the plate tectonics regime, the Camels Hump group, collectively, appears to represent the type of sedimentary and volcanic accumulation that can be observed on the outer edges of the continents today, on the so-called continental slopes and rises. These, however, were deposited more than 500 million years ago when the continent was

Small gorge in Pinnacle formation volcanic member at Enosburg Falls. Layers dip steeply left.

smaller and the coast did not extend as far east as it does at present. Most of the sediments were derived from the wearing down of the ancestral Adirondacks to the west. All were crumpled and metamorphosed much later during the Taconian and Acadian mountain-building events.

Our route follows the Missisquoi River along Vermont 105 between Sheldon Junction (1 mile southeast of East Highgate) and East Berkshire (13 miles). The valley is fairly broad and open, with much farmland. Clearly visible on the sides are remarkably well-preserved river terraces that descend in stair-step arrangement to the river. They formed at the end of the last ice age. When the waters of Lake Vermont covered this section of the state, the valley was filled from one bedrock wall to the other with water at least to the level of the highest terrace, and thick sediment deposits accumulated on the valley floor. When Lake Vermont drained, the level of the water in the Champlain basin dropped about to its present level. This initiated a long period of erosion during which the Missisquoi River cut downward into the valley fill as it meandered back and forth across the valley, leaving terrace remnants here and there where the sediment removal was incomplete. Later, in response to a worldwide rising sea level, the valley flooded again and Champlain Sea sediments were deposited on top of the remaining valley fill, but only on the lowest terraces because the water level was now much lower. The newer deposits are easily distinguished from those of Lake Vermont by their

78

242

abundant fossil shells. None has been found on the upper terraces. Nearly all of the small tributaries to the Missisquoi are entrenched into the valley sides. Their downcutting was initiated by the downcutting of the master stream.

Bedrock outcrops are fairly numerous in the farm fields and pastures on the Missisquoi Valley sides. The river has cut a small gorge at Enosburg Falls through the greenish, chlorite-bearing Tibbit Hill volcanic member of the Pinnacle formation.

The solitary high peak visible to the east from many places in the Missisquoi Valley is Jay Peak. Between East Berkshire and Jay (20 miles), you leave the Missisquoi and follow its tributaries, Trout and Jay brooks, along Vermont 118 and 242 around the south and east sides of the mountain. The mountain is a lone survivor of erosion of a once much higher, and presumably, less rugged landscape, now surrounded by a surface of low relief. In this case and throughout Vermont, New Hampshire, and all of the northeast, the landscape reduction is the product of both stream and glacial erosion.

Jay Peak is also an interesting study in alpine glacial erosion. The summit area has been deeply scalloped by small glaciers that coursed down the existing stream valleys in the early, and possibly also late, stages of continental glaciation, much as they do in the Alps today, but on a smaller scale. The tell-tale features are steep-sided, bowl-shaped amphitheaters at the valley heads and a valley downstream that shows a broadly gouged cross section. These are quite evident from the road near Montgomery.

A number of high peaks in this region reveal asymmetrical profiles when viewed from the east or west, with a relatively gentle north slope and steep, even cliffy, south slope. This is a common feature of glacial erosion called a sheepback, produced when overriding ice, in this case moving generally southward, scraped and polished the north sides of bedrock knobs, and plucked rock away from the south sides where it was unsupported. One such peak is prominently profiled at Hazens Notch southeast of Montgomery Center. Montgomery Center, incidentally, lies at about the limit of the Missisquoi arm of glacial Lake Vermont.

Between Montgomery Center and Jay (13 miles), Vermont 242 crosses the Gilpin Mountain—Jay Peak Pass with a rise of 1800 feet on the west and 1300 feet on the east side. From the pass you have a good view to the northeast of the piedmont, or New England upland, section of Vermont, the region east of the Green Mountains that is deeply eroded and mantled with sediments from the mountains. From this vantage point, and in the backward view from Jay Valley, the isolation of Jay Peak is more apparent.

78
105
118
242

The village of Jay is just west of the eastern limit of the Camels Hump group. A distinctive feature along the boundary in this region and over nearly the full length of the state is numerous pods and slivers of serpentinite, a dark green rock composed of the mineral serpentine, and other closely associated dark-colored rocks. This same zone extends north to the large asbestos deposits of Thetford Mines in Quebec. These dark rocks appear to be altered slivers of oceanic crust that were carried upward along faults in the early stages of the Taconian mountain-building event. Vermont 105 crosses a small serpentinite sliver about two miles east of North Troy. Between North Troy and Newport, the route crosses north-south bands of schist of Cambrian to Devonian age. Some geologists now suspect that these bands really belong to multiple thrust sheets that were stacked against each other in Taconian time.

The route near Newport is punctuated with beautiful views of Lake Memphremagog. The lake basin was the site of a large, high-level lake during Wisconsin deglaciation, higher than Lake Vermont. On the valley sides near Newport are remnants of beach strands, wave-cut benches, and other shoreline features. The highest is about 320 feet above the present lake; lake sediments veneer the slopes below, but not above, that level. The same high level lake also extended an arm into the upper Missisquoi Valley around Jay. As in the case of Lake Vermont shorelines, these have been uplifted as a result of glacial rebound. The shorelines and lake deposits here can be traced

Xenolith of early Devonian Waits River schist in granite of Derby pluton at exit 28 of Interstate 91. Schists here are contact-metamorphosed to sillimanite zone.

78 105 118 242

for many miles south along the Black River Valley to the Lamoille Valley. The valley also contains abundant deposits of till and glacial outwash.

The section of Vermont from Newport eastward to the Connecticut River is part of a belt of early Devonian rocks that continues southward into Massachusetts. The rocks belong to the Waits River and Gile Mountain formations. The belt is also punctured by numerous large and small bodies of granite and related rocks that belong to the middle Devonian New Hampshire plutonic series. There are dramatic roadcuts at exit 2 of Interstate 91, two miles east of Newport, at the edge of the Derby granite pluton. The white granite intrudes dark schists of the "country rock," penetrates along fractures and between layers, and engulfs numerous fragments of broken schist, or "xenoliths." Near contacts, the intruded rocks are also more intensely metamorphosed by the heat of the intrusion. Many on the north edge of the pluton contain the high temperature, metamorphic mineral sillimanite, a product of the contact metamorphism.

Vermont 100: The Talc Road

Vermont 100 is one of the state's lesser-known, yet most scenic roads. It stretches for over 200 miles up the middle of the state within or very near the serpentine belt discussed in the Plate Tectonics section. Nearly all of the many serpentine bodies are elongated parallel to the north-to-south geological trend, and serpentine and host rocks are highly sheared. The belt consists of metamorphosed sedimentary and volcanic deposits that were scraped from the ocean floor on the landward side of the Bronson Hill island arc in the early stages of the Taconian mountain-building event, transported upward and westward, and plastered onto the continental margin. The serpentine slivers are pieces of dark rock ripped from the oceanic crust and carried up along thrust faults to their present level. They are really altered dunite, an olivine rock; peridotite, olivine-pyroxene rock; or gabbro, pyroxene-feldspar rock, some of which survived complete serpentinization. Nearly all of the slivers contain talc, an end-product of low-grade metamorphism.

Serpentine has been widely quarried for verde antique, one of the most beautiful decorative stones used in many Vermont buildings. Rough-cut and polished slabs of this deep green rock are stacked alongside cut marble in the yards of Vermont Marble Company at Proctor. Some of the minerals associated with the serpentine are asbestos, the fibrous form of serpentine, talc, actinolite, tremolite, magnetite, chromite, magnesite, dolomite, calcite, pyrite, pyrrhotite, bornite, chalcopyrite, graphite, limonite, and nickel ore minerals. There are numerous other collectable minerals in the host rocks, most of which are metasedimentary schists of the Camels Hump group.

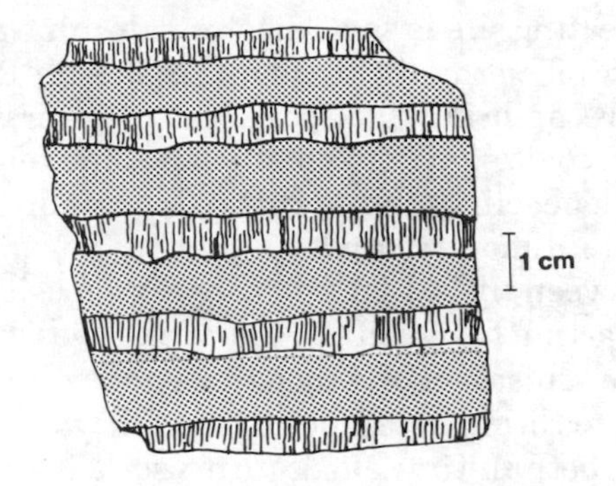

Chrysotile, the asbestiform serpentine, often occurs in veins like this in dense antigorite serpentine.

Vermont 100: Wilmington—Hancock

100 mi. / 160 km.

see map page 76

The village of Wilmington lies near Harriman Reservoir on the Deerfield River at the junction of Vermont 100 and Vermont 9, the Molly Stark Highway. It's the only substantial settlement on Vermont 9 between Brattleboro and Bennington. Over most of the way between Wilmington and Pittsfield (84 miles), the road follows the eastern edge of the exposed Precambrian core gneisses of the Green Mountains. The exceptions are the 31-mile section between East Jamaica and Ludlow and short spans just north of Wilmington and southwest of Ludlow, where the road lies within the Precambrian terrane. Roadcuts expose grayish Precambrian gneiss along each of these road sections.

The largest body of the Vermont serpentine belt, the Dover body, is exposed in cuts at East Dover, 5 miles from West Dover on the road to West Dummerston. The road passes through the middle of the body, which is 3½ miles long from north to south and a mile wide. Dense serpentine and asbestos make up most of the roadcut, but north of the highway, the body contains talc and partially serpentinized dunite. Some apple-green serpentine is thought to be a nickel-bearing variety called garnierite, which forms by alteration of nickel-rich peridotite. Cuts in Precambrian rock between East Jamaica and Ludlow (31 miles) reveal highly sheared, micaceous gneisses that locally contain small knots of reddish garnet. Some cuts, however, expose massively-bedded quartzites and micaceous quartzites that are widely distributed among and interlayered with the gneisses in this section of the Green Mountains.

An excellent mineral collecting site is the Carleton talc mine about 12 miles east of Londonderry off Vermont 11, near the Chester Reservoir, and very close to the Precambrian gneisses of the Chester dome. The host rock is quartz-muscovite-garnet schist with some feldspar and chlorite, and minor sphene, clinozoisite, rutile, tourmaline, and the metallic minerals pyrite, limonite, and magnetite. Several interesting mineral zones between the schist and the talc deposit were produced by chemical interaction between the two rock types during metamorphism. Next to the schist is a one-foot-thick zone of coarse-grained, dark biotite mica with minor amounts of yellowish-green epidote, apatite and zircon, both in tiny elongate crystals. Next is a thin zone of soft green chlorite, then an eight-inch zone of green, needle-like prisms of actinolite. The core rock consists of talc, dolomite, and magnesite, which reflect the high magnesium content of the rock, a characteristic of all serpentine masses. In addition to the above minerals, some of the best specimens are magnetite which occurs in octahedral crystals up to ½ inch in diameter and cubes of shiny pyrite, fool's gold, as much as one inch across. Both minerals occur in a matrix of chlorite soft enough to permit free growth of such well-formed crystals.

There is a different type of mineral-collecting site north of Ludlow on Vermont 103, about one mile from the Vermont 100 intersection. This is a long roadcut opposite a parking area that exposes a number of rock types along the contact between the Precambrian core rocks and the enclosing, younger schists. Included are several different gneisses, quartzites, pegmatites, marbles, mica schists, and calc-silicates, rocks largely composed of calcium-silicate minerals and often grayish green in color. Good specimens of talc, calcite, diopside, pyrite, and tourmaline may be found here.

Another collecting site is one mile east of Tyson (5 miles north of Ludlow) along the road to South Reading, in cuts on both sides of the road and outcrops in the bed of Kingman Brook. The rock is quartz-muscovite-feldspar-biotite-chlorite schist of the Camels Hump group. It contains good specimens of garnet, magnetite, and chloritoid. The latter mineral, which is widespread in the low- to medium-grade metamorphic rocks of Vermont, occurs here in exceptionally fine, large, rectangular crystals up to ½ inch in diameter.

The area around Plymouth (5 miles north of Tyson) is one of many parts of Vermont that has yielded gold. The gold occurs as concentrations of heavy grains called placers in the stream beds that can be recovered by panning or sluicing. Small quantities may still be found this way.

Between Plymouth and West Bridgewater (5 miles) are several

Precambrian quartzite within the Green Mountain massif near Sherburne Pass.

roadcuts in buff dolomite and phyllites of the Tyson formation, the basal part of the Camels Hump group. Nearly all of the rock strata along the eastern margin of the Green Mountains dip steeply eastward away from the Precambrian core.

The 4-mile segment of highway between West Bridgewater (5 miles north of Plymouth Union) and Sherburne Center follows a glacially-gouged section of the Ottauquechee River valley. The road goes right along the eastern boundary of the Green Mountain core, and all of the rocks on the valley floor are of Cambrian age. The unconformity between the two groups of rocks represents an erosional gap in the geologic record of more than 500 million years.

From Sherburne Center the road climbs westward for two miles to Sherburne Pass (2190 feet), then turns north and makes a winding, woodsy descent along Tweed River to its junction with White River at Stockbridge, 12 miles from the pass.

Between Stockbridge and Hancock (12 miles) you follow the White River. Some wide sections of the valley have flat floors, the result of glacial-lake sedimentation. Occasional terraces are similar to those of the Connecticut Valley. This lovely traverse is blessed with many unobstructed views of the bordering mountains. Roadcuts expose greenish, rusty-weathering, chlorite-bearing schists or gneisses of

100

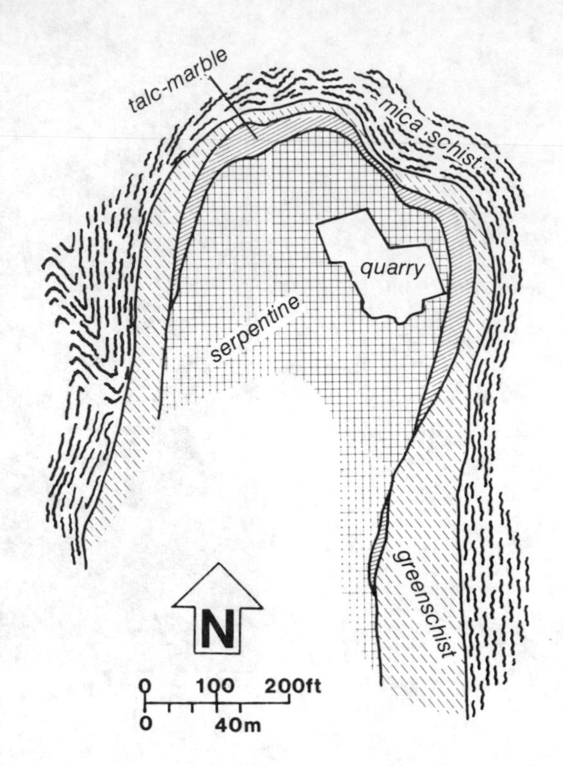

Map of Vermont Marble Co. Verde Antique (serpentine) Quarry at Rochester, showing inner talc-marble, and outer greenschist contact zones. Strong shearing in the enclosing schist generally follows the contact. —From P. H. Osberg.

the Camels Hump group far from the Precambrian contact.

Talcville, 7 miles north of Stockbridge, is a former talc mining community. The only verde antique quarry now active in the state is the one near Rochester. The core of the serpentine body is rimmed by talc-carbonate rock. This, in turn, is surrounded by feldspar-epidote-carbonate schist that is probably a reaction zone between the serpentine and the enveloping mica schist.

Vermont 100: Hancock—Troy

80 mi. / 128 km.

Hancock is at the junction of the Hancock branch with the main branch of White River. Vermont 100 follows the main branch upstream to Granville (4 miles) at the junction of two more tributaries, Alder Meadow Brook and Kendall Brook. Moss Glen Falls is a major scenic attraction between Granville and Warren (10 miles). You cross the divide in this section between the south-flowing White River, tributary to the Connecticut, and north-flowing Mad River, tributary to the west-flowing Winooski which, in turn, discharges into Lake Champlain. The Mad River tumbles through a steep rocky gorge south of Warren exposing a large expanse of greenish Camels Hump schist. Lincoln Gap (2424 feet), a narrow rocky pass in the high mountains west of Warren, was cut by streams and glacial ice. It is similar to Smugglers and Hazens notches, Middlebury, Mount Holly and Brandon gaps, and Sherburne Pass. Wisconsin ice that moved across the Green Mountains from the northwest crowded thick tongues through these passes and gouged them into deep narrow slots.

The Mad River talc mine and Duxbury serpentine quarry are in different parts of the same serpentine body near South Duxbury (5 miles north of Waitsfield). Minerals at these two localities include massive serpentine, talc, green actinolite, white fibrous tremolite, magnetite in octahedrons up to ½ inch across, and rare disseminated grains of chromite, a chromium mineral that also forms octahedral crystals. The Waterbury talc mine is an abandoned underground mining operation just south of US 2, southeast of Waterbury where collectors find typical serpentine-related minerals in the dumps.

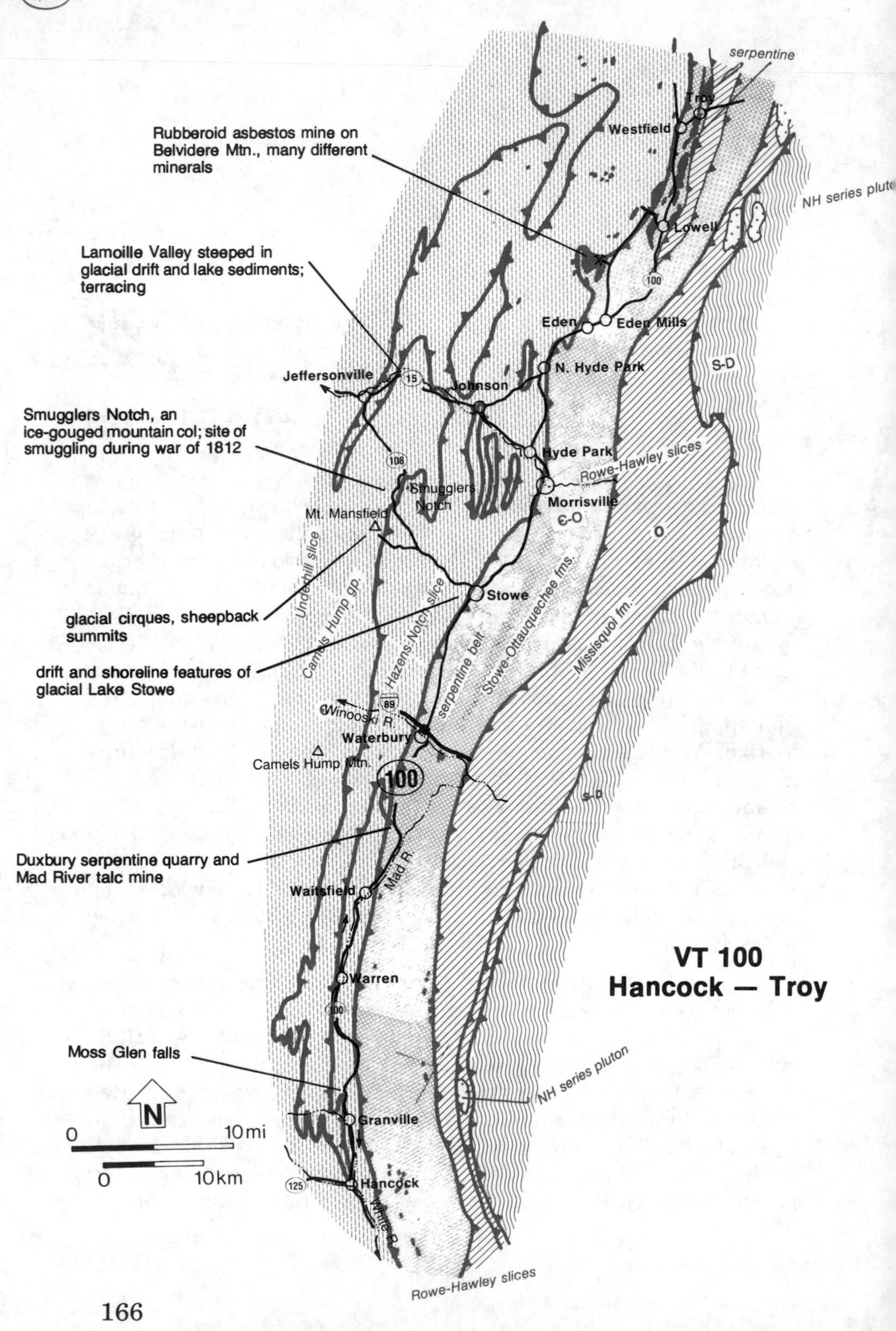

serpentine
Troy
Westfield
Rubberoid asbestos mine on Belvidere Mtn., many different minerals
NH series pluton
Lowell
100
Lamoille Valley steeped in glacial drift and lake sediments; terracing
Eden
Eden Mills
N. Hyde Park
S-D
Jeffersonville
15
Johnson
Smugglers Notch, an ice-gouged mountain col; site of smuggling during war of 1812
Hyde Park
Rowe-Hawley slices
108
Smugglers Notch
Morrisville
Mt. Mansfield
Є-O
O
Underhill slice
Camels Hump gp.
Hazens Notch slice
Stowe
Stowe-Ottauquechee fms.
glacial cirques, sheepback summits
drift and shoreline features of glacial Lake Stowe
serpentine belt
Missisquoi fm.
89
Winooski R.
Waterbury
Camels Hump Mtn.
100
S-D
Duxbury serpentine quarry and Mad River talc mine
Mad R.
Waitsfield
VT 100
Hancock — Troy
Warren
100
Moss Glen falls
NH series pluton
N
Granville
0
10 mi
0
10 km
125
Hancock
White R.
Rowe-Hawley slices

Excellent views of the Winooski Valley open along this southern approach to Waterbury. The valley is one of three major water gaps through the Green Mountains that served as channels for advance and retreat of Wisconsin ice. Thick deposits of glacial drift and glacial lake sediments, including those of early high level lakes and the late Lake Vermont, underlie the broad, terraced valley floor. Enormous roadcuts at the Waterbury-Stowe interchange to Interstate 89 display biotite schists with nearly vertical layering near the contact between the very similar Ottauquechee and Stowe formations. Vermont 100 follows this contact most of the way to Troy (48 miles).

The main geologic features around Stowe are of glacial origin. The wide open valley is filled to considerable depth with glacial drift and glacial lake sediments deposited during the slow recession of Wisconsin ice across the Green Mountains. The valley sides abound with lakeshore erosional and depositional features. While the Lamoille, north, and Winooski, south, drainage routes across the mountains were still completely blocked by ice, Lake Stowe ponded in the valley. After further ice recession, it became part of a somewhat lower lake that flooded the Winooski, Stowe, and Lamoille valleys, connecting with the latter at Morrisville. Still later, Lake Vermont reached across the mountains through the Winooski and Lamoille valleys, and covered the same region at an even lower level, maintaining the Stowe Valley connection. In all of these stages, long arms of the lakes reached up tributary valleys. The lake levels, in all cases, were controlled by outlets. Progressively lower outlets opened as ice blockages melted away from them, permitting water levels to drop. Shoreline features are only well-developed at levels that remained static for long periods. The highest beach deposits around Stowe and Morrisville are at 1200 feet, about 500-600 feet above the present valley floor.

Two mineral-collecting sites are east of Morrisville, both in rocks of the Stowe formation. The Morrisville lead-zinc prospect is northeast of the village, directly north of the Vermont 15-15A intersection. Pyrite, silvery cubes of galena, sparkling yellowish sphalerite, calcite, and barite occur in cross-cutting veins in mica-chlorite-feldspar schist. Dense, white barite is the most abundant mineral. The Wolcott copper mines are on Toothacher Hill north of Vermont 15, about 2 miles east of the Morrisville prospect. Pyrite, magnetite, silvery pyrrhotite, golden yellow chalcopyrite, sphalerite, and galena occur in greenschist.

The Johnson talc mine is in the hills west of North Hyde Park, 8 miles north of Morrisville. Mine dumps there contain pyrite, magnetite, dolomite, magnesite, and less common pyrrhotite, colorful bornite, chalcopyrite, graphite, ilmenite, and nickel ore, in addition to the

usual serpentine and talc. The serpentine body is 3500 feet long from north to south and 200 feet wide.

Another good mineral collecting site is the Ruberoid asbestos mine on Belvidere Mountain north of Eden Mills, 5 miles northeast of North Hyde Park, or southwest of Lowell, 8 miles north of Eden Mills. Secondary roads from both villages lead to the mine. The quarries are in a large mass of serpentinized dunite and peridotite, surrounded by schists, gneisses, quartzites, and amphibolites of the Camels Hump group. Some 32 mineral species have been found here, many specimens of museum quality. Magnetite, chromite, and olivine are the primary minerals formed with the dunite or peridotite. Talc, serpentine, asbestos, brucite, chlorite, siderite, biotite, calcite, magnesite, pyroaurite, and dolomite are alteration products from reaction of the primary minerals with water and carbon dioxide. Diopside, clinozoisite, grossular, idocrase, sphene, zoisite, graphite, apatite, and rare green uvarovite garnet, all formed by reaction between limestone or dolostone inclusions and the original dunite and peridotite magma. Bornite, chalcopyrite, pyrite, and chalcocite formed from later solutions that brought copper into the rock. Finally, artinite, albite feldspar, prehnite, stilbite, and some of the calcite, clinozoisite, and zoisite formed as very late fracture fillings deposited from hot water solutions.

More serpentine bodies concentrate near Vermont 100 between Lowell and Troy (8 miles) than in any other part of the Vermont serpentine belt. But there are no outstanding mineral collecting sites in this section.

The flat terrain around Westfield, Troy, Jay, and North Troy is, once again, the sediment plain of proglacial lakes, including high-level lakes of the Memphremagog basin and Lake Vermont, which extended a long arm into Canada and then southward to here.

Mount Mansfield from Vermont 108 near Stowe. Clearly defined from left to right are the "forehead," "nose" with towers, "chin," and "Adam's apple." Wells Brook in foreground cuts into Lake Stowe sediments.

Mount Mansfield, Vt. Toll Road

8 mi. / 13 km. round trip

see map page 166

At 4393 feet above sea level, Mount Mansfield is the highest point in Vermont. The total rise from the toll house to the end of the toll road near its summit is about 2400 feet.

The Indians called Mansfield Moze-O-Be-Wadso: mountain with Head like a Moose. White men likened it to the head of a man in repose, with bumps along the summit ridge which, when seen from the west or east clearly define the forehead, nose, chin, and Adam's apple. On clear days, there are great views from the nose, a short climb from the parking lot at the end of the road. To the west, nearly all of Lake Champlain is visible, with the shadowy profiles of the Adirondacks on the far side. The lowlands of the Champlain basin contrast more strikingly with the Green Mountains than when seen from ground level. To the east is broad Stowe Valley, flanked by the Worcester Mountains and the New England uplands beyond, with their scattered high peaks; the great domical mass of the White Mountains looms in the far distance. The high, solitary summit of Mount Ascutney is almost directly south; beyond and a little east of it you can almost see the New Hampshire coast. The Green Mountain crest marches along for many miles from north to south. Camels Hump (4083 feet) thrusts its pointed peak to the sky 15 miles to the south. It is carved from greenschists nearly identical to those of Mount Mansfield, and also provides grandstand views of the surrounding country.

The lower three miles of the toll road are in the trees; the last mile ascends the south rim of the ski bowl with excellent overviews of the east side of the mountain and Smugglers Notch. The notch is a very narrow rocky pass that divides the watersheds of the north-flowing Brewster River, tributary to the Lamoille River, and the south-flowing West Branch River, tributary to the Winooski River. Streams originally carved it, but glaciers opened it wider. The bedrock is greenish quartz-muscovite-chlorite schist of the Cambrian Camels Hump group. The rock layers are extensively crumpled into small chevron folds with numerous thin, white quartz interlayers and lenses that weather in relief. Many exposures have a pearly luster because of the way the fine-grained muscovite mica reflects the light. The rocks are in a gigantic thrust sheet, called the Underhill slice, which transported them far to the west of their place of origin during Taconian mountain-building.

Glacial erosion is largely responsible for the sculpturing of Mount Mansfield. The peak stands at the edge of the Green Mountains, virtually unprotected by other mountains or foothills on its northwest side. During Wisconsin and earlier glaciations, it withstood the direct impact of ice advancing from the northwest gouging and rounding the existing stream valleys on that side. The summits were eventually reshaped into rounded knobs like oversized sheepbacks when the ice sheet finally overrode them. In fact, the whole mountain mass became asymmetrical, its southeastern side considerably steepened. Ice that jammed into Smugglers Notch was particularly effective in deepening it and steepening its walls. The ski bowl is the largest and most cirque-like valley on the mountain, suggesting that it may have held a small alpine glacier for a while, either in the early or late stages of Wisconsin glaciation, or both.

Ski bowl and Smugglers Notch from the Mount Mansfield auto road.

Smugglers Notch from southeast side.

Vermont 108: Stowe—Jeffersonville through Smuggler's Notch

18 mi. / 29 km.

see map page 166

Thick deposits of glacial drift and lake sediments, mostly of sandy composition, underlie the broad, flat floor of Stowe Valley. Water covered this valley for a very long period during Wisconsin deglaciation of the Green Mountains and the Lake Champlain basin. The submersion began as soon as the ice left the valley and the glacier front still presented an impenetrable barrier against discharge of the meltwater to the west. The water could not escape through the Lamoille and Winooski valleys because they were completely blocked by ice. A high level glacial lake formed in the Stowe Valley at an elevation of about 1200 feet. Shoreline features of this lake, such as beach gravels, sand bars and wave-cut benches, are well-preserved around Stowe today. A succession of lower high-level lakes followed as the ice margin receded and opened new, lower outlets. After the Lamoille Valley opened, a long arm of Lake Vermont reached through it from the Champlain basin to here.

On Vermont 108, the West Branch River valley between Stowe and the Mount Mansfield gondola (8 miles) is mostly open, with good

Smugglers Notch southwest cliff which is particularly prone to landsliding. Early Cambrian Camels Hump group greenish schist.

views of the mountains all around. Most impressive is Mount Mansfield itself, with its three bumps—the nose with transmission towers on it, the chin, the true summit, and the Adam's apple. There are almost no bedrock outcrops. The road banks consist of gravelly glacial drift.

The last 2 miles from the gondola to the notch are very steep. The approach is punctuated with glimpses of the rocky cliffs of the pass. The notch is a shadowy fantasyland of incredibly high cliffs and huge boulders fallen from them, all overgrown with green jungle. One such talus block, called King Rock, bears a sign noting that it weighs 6000 tons, and that it tumbled from the cliff in 1910. The cliffs of the west or Mount Mansfield side of the pass, are extremely ragged, with many overhangs, and are, perhaps, more prone to rock fall than those of the east side. Three recent slides are visible in the notch area; one on the east near the beginning of the steep climb; one distinguished by a strip of pale, unweathered rock on the west side of the pass; and another marked by a fan of uprooted trees and debris adjacent to the east side of the road near the pass. The rocks of the notch, and in fact, all of the nearby mountains, belong to the Camels Hump group of early Cambrian age. Its attractive green color comes from the soft, micaceous, metamorphic mineral chlorite. Layering in the rock is highly contorted and well-defined by interlayers and lenses of hard quartzite that weather in relief. Recent studies indicate that these rocks originated far east of their present position and were transported here, in a gigantic thrust sheet, called the Underhill slice during Taconian mountain-building. Folding and metamorphism happened at the same time as fault movement.

Smugglers Notch is so-called because forbidden goods, including cattle, passed through it during the War of 1812 at the time of

Jefferson's Embargo Act, which prohibited trade with Canada. At that time, there was only a trail and horse path through the pass, and the smugglers' activities were remote from the watchful eyes of the authorities. The first carriage road was built in 1894, and this was upgraded to an automobile road in 1918. Between the notch and Jeffersonville (8 miles), the road descends nearly 1700 feet along the Brewster River to the Lamoille Valley. Most of the drop occurs in the first three miles. The lower five miles is fairly open valley from which you can see Mount Mansfield (4393 feet) west of the pass. The several peaks of the Sterling Range on the east side, from north to south, are White Face Mountain (3715 feet), Morse Mountain (3385 feet), Madonna Peak (3640 feet), and Spruce Peak (3320 feet). The latter four peaks are just bumps on a continuous arcuate ridge that, together with the northward extension of the Mount Mansfield summit ridge forms a funnel-like entrace to the notch open to the north. This topography probably jammed ice into the notch thus contributing to its spectacular sculpture. The ice advanced on the Green Mountains from the northwest and crowded against the western slope until it overtopped the crest of the range. Here it very likely piled up in the funnel and spilled out through the notch well before it crept over the mountain tops.

Jeffersonville lies on a broad section of the Lamoille Valley that stretches for 20 miles from Fairfax Falls to Hyde Park, where it joins with north-south-trending Stowe Valley. The valley floor here is a flat flood plain developed on glacial drift and Lake Vermont deposits, and the river meanders widely over it nearly from wall to wall. High glacial lake deposits are found only in the upper valley east of here, where they are continuous with those of Stowe Valley. There are many sandy terraces on the sides of this 20-mile stretch of river, some of which extend far up into the northern tributaries.

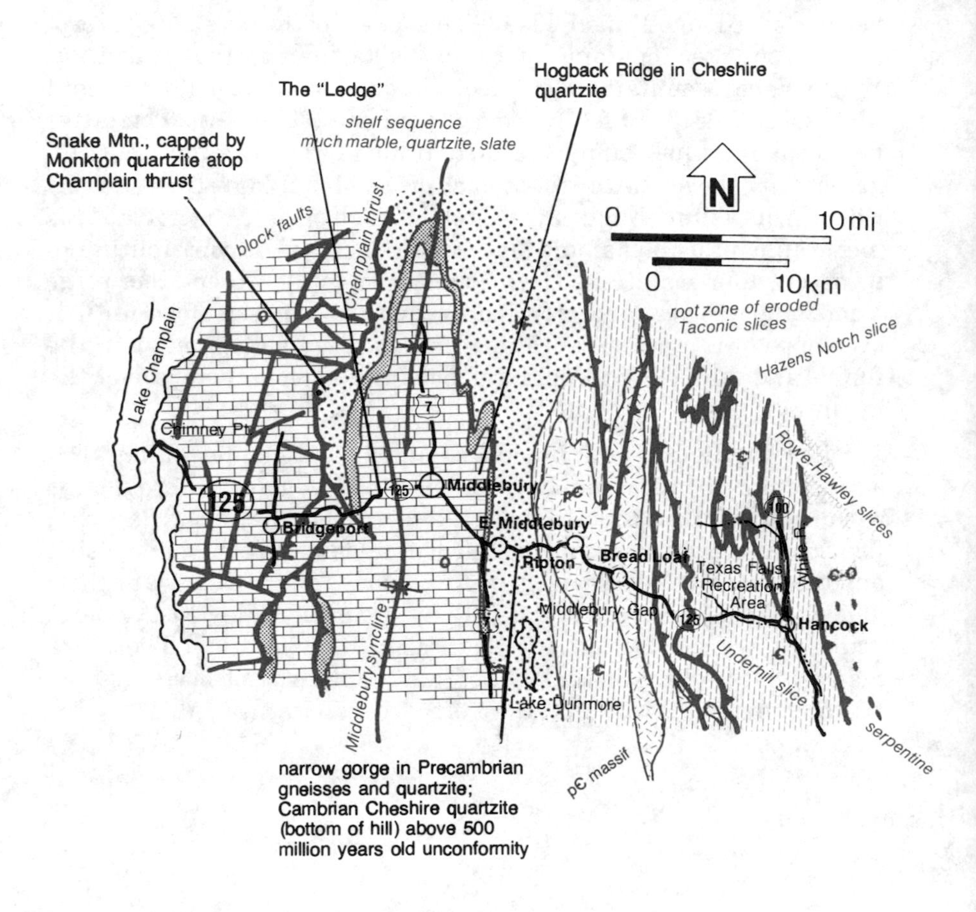

VT 125
Hancock — Chimney Point

Vermont 125: Hancock—Chimney Point

37 mi. / 59 km.

Between Hancock and Bread Loaf (10 miles), the road follows the Hancock branch of White River past Texas Falls Recreation Area. About midway, you pass through Middlebury Gap (2149 feet), one of several Green Mountain passes that were reshaped by ice. Undoubtedly, all were initially carved by streams before Wisconsin glaciation, but when the ice massed against the west side of the mountain front, these gaps permitted early movement through them to the east side of the range. The tongue-like protrusions probably continued even as the glacier thickened and covered the mountains completely.

The pass is in the erosionally-truncated "root zone" of the Taconic slices, and the road west of the pass crosses over the thrust faults along which they moved. West of here, erosion has completely destroyed the slices, but farther south they are preserved in the Taconic klippe.

Between Bread Loaf and East Middlebury, you span the Precambrian Lincoln massif, revealed in a few roadcuts as medium gray, fine-grained gneiss with strong layering and whitish, massively-bedded and highly fractured quartzite. The rocks are similar to those of the large Green Mountain massif to the south. The upper part of the deep narrow gorge between Ripton and East Middlebury (4 miles) is in this quartzite; and the lower part, as for example at the bridge at the lower end of the gorge, is in thin-bedded early Cambrian Cheshire quartzite, which forms a persistent ridge along the mountain front, the Hogback Range.

East Middlebury is on the east limb, or side, of the Middlebury syncline, a large downfold in the metamorphosed Cambrian and Ordovician sedimentary strata that underlie the region between the Green Mountains and the Champlain thrust fault—therefore, in the Champlain thrust slice. In map view, bands of rock, particularly of marbles formed from limestones and dolostones, define the fold as an elongate wedge that points northward, with the apex near

Vergennes. These rocks were originally deposited on the same Precambrian gneisses and quartzites exposed in the Lincoln massif. The Precambrian contact surface is a profound unconformity, representing an erosional gap of about 500 million years; it is the same boundary that separates the Cambrian and Precambrian quartzites at East Middlebury. The relief.between the unconformity at the bottom of the syncline and the top of the Green Mountains is at least three miles, and perhaps as much as six miles! When the original sediments accumulated on top of it, the Precambrian erosional surface was nearly horizontal continental shelf.

Marbles and quartzites of this region have been widely used in building and are well-displayed on the campus of Middlebury College. The prominent hill directly north of Middlebury is carved from quartzose marble of the middle Ordovician Beekmantown group, formed from original sandy limestone. Numerous outcrops of Ordovician limestones are at the core of the syncline in the low hilly country between the village and "the Ledge," located 2½ miles east on Vermont 125. All of the hills here are elongate from north to south and resemble drumlins, but most appear to be streamlined rock knobs called sheepbacks that have been scraped smooth by overriding ice. At the Ledge, the road steeply descends about 300 feet to cross Lemon Fair River. From here westward is open farmland with splendid views of both the Green Mountains and the Adirondacks.

You cross the Champlain thrust fault between the Ledge and Bridport (5 miles). The one high hill north of the road is Snake Mountain (1282 feet), its summit supported by hard Cambrian Monkton quartzite that overlies the fault, visible in the cliff of the west face.

If we consider bedrock structure, only the flat country west of the thrust is Champlain lowland, for the fault divides highly deformed, out-of-place rocks of the thrust slice from in-place rocks that are only mildly deformed. Rocks to the west have also been down-dropped, creating the basin that holds Lake Champlain. Physiographically, however, the Champlain lowland is considered to extend to the Green Mountains. The lowland topography throughout was severely reduced by glacial erosion, then buried under thick deposits of drift and Lake Vermont and Champlain Sea sediments. The sedimentary cover is thinner now than originally, for stream erosion following postglacial uplift has removed much of it.

Middle Ordovician Trenton group shales and limestones underlie the flatland between Bridport and Chimney Point (8 miles). The beds lie more or less flat but several block faults like those of the east edge of the nearby Adirondacks break them.

Rocky coast at Rye Beach, New Hampshire, in schists and gneisses of the Silurian Rye formation.

New Hampshire 1A: The New Hampshire Seacoast

15 mi. / 24 km.

see map page 202

The northern two-thirds of the short New Hampshire coastline consists of ragged promontories that separate pocket beaches. The rocks are tough, erosion-resistant schists and gneisses of the Rye formation that occupy the core of the Rye anticline. The core region is shaped like a wedge about 4 miles wide in a northwest direction and 12 miles long, which points southwestward where the fold dives under younger rock layers. The open end of the wedge is at Portsmouth Bay, the point at Hampton Falls.

Where exposed on the shore, rock layers of the Rye formation stand on end and display abundant, small-scale deformation. The rocks are principally highly metamorphosed sediments and volcanic material with cross-cut and interlayered pegmatites. Nearly all were highly sheared and then healed by later metamorphism. This is most apparent in the coarse-grained, whitish pegmatites where shear surfaces weave in and out among eye-shaped fragments of white feldspar. Rock layers are locally bent into small folds or offset by minor faults. Numerous dikes and sills of dark basalt are up to about 6 feet wide; these are not metamorphosed and, therefore, are considerably younger than the host rocks. To explain, many of the host rocks here contain the minerals garnet, staurolite, or sillimanite, which crystallized many miles below the surface. The basalt dikes, on the other hand, formed with intrusion of magma near the surface. Thus, miles of rock overburden had to be removed by erosion before the basalt was intruded—and that takes a long time. It can safely be assumed that the land surface is not much lower now than it was at the time of the intrusions.

The inland part of the Rye formation near the axis and in the northwest limb is intruded by several small bodies of granite of unknown affinity that may belong to the New Hampshire plutonic series. Similar rocks surface in the Isles of Shoals about 6 miles offshore.

Metamorphism and contemporaneous deformation of the Rye formation appears to be entirely Acadian. No conclusive evidence has yet been found that the rocks were imprinted with the late metamorphism of the Alleghenian mountain-building event. Even so, the rocks here seem to be more intensely sheared and granulated than most of those farther inland.

The southern part of the New Hampshire coast is less rocky and more sandy. This is partly due to the lesser resistance of the slate bedrock belonging to the Eliot formation and partly to the drowning of the mouth of Hampton Falls River which forms Hampton Harbor. Farther inland, Interstate 95 parallels the shore and traverses the same rock layers on the western limb of the Rye anticline with several good roadcuts.

Sheared and folded schist and pegmatite in Silurian Rye formation, Rye Beach, New Hampshire.

New Hampshire 9: Concord, N. H.— Brattleboro, Vt.

57 mi. / 91 km.

Franklin Pierce Highway

This section of New Hampshire 9 skirts the northern edge of one of the major drumlin belts of the United States with more than 3000 drumlins, most of them in Massachusetts. Many of those in New Hampshire appear to be rock drumlins, which have a streamlined shape similar to till drumlins, but consist of bedrock knobs smeared with glacially molded till. However, only a small portion of the rounded knobs in this hilly, rolling country are drumlins, with or without rock cores. The rest are simply products of uneven stream erosion of the land followed by glacial scour during the Ice Age.

The dominant bedrock exposed in the few roadcuts between exit 5 of Interstate 89 and Keene (43 miles) belongs to the early Devonian Littleton formation and the late Devonian New Hampshire series plutons that intrude the Littleton. The Littleton rocks are schists and gneisses and micaceous quartzites. They contain the mineral sillimanite, which indicates high-grade metamorphism. Many of the old roadcuts have grown rusty through oxidation of iron-bearing minerals. Inasmuch as the rocks are dated as early Devonian by fossils, the metamorphism is considered to be a product of middle-to-late Devonian Acadian mountain-building. Some of the gneisses contain many very pale, granitic layers, veins, and lenses. The magma probably formed through partial melting during the extreme metamorphism of the host rock. Those rocks full of small bodies of granite are called migmatite, a descriptive term for an intimate mixture of igneous and metamorphic components; they are very common in regions of high-grade metamorphism. Several roadcuts in the Littleton formation also contain dikes of extemely coarse white pegmatite or very fine-

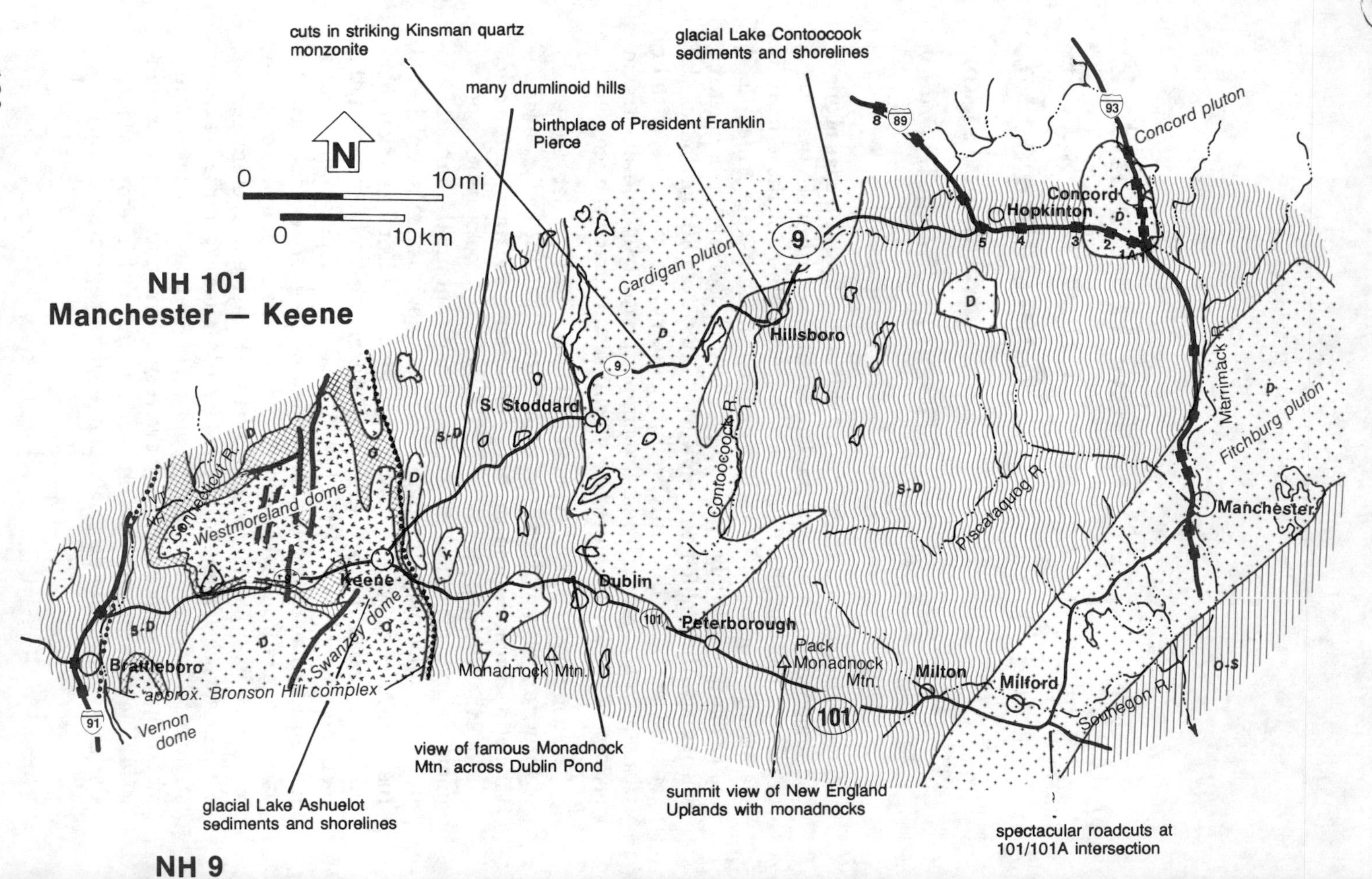
9
cuts in striking Kinsman quartz monzonite
many drumlinoid hills
birthplace of President Franklin Pierce
glacial Lake Contoocook sediments and shorelines
0
10 mi
0
10 km
N
NH 101
Manchester — Keene
Concord pluton
Concord
Hopkinton
93
89
8
5
4
3
2
1A
D
Cardigan pluton
9
Hillsboro
S. Stoddard
S-D
Merrimack R.
Fitchburg pluton
Manchester
Contoocook R.
Piscataquog R.
Westmoreland dome
Connecticut R.
VT
Keene
Swanzey dome
Dublin
101
Peterborough
Pack Monadnock Mtn.
Monadnock Mtn.
Milton
Milford
Souhegon R.
O-S
Brattleboro
approx. Bronson Hill complex
91
Vernon dome
view of famous Monadnock Mtn. across Dublin Pond
summit view of New England Uplands with monadnocks
glacial Lake Ashuelot sediments and shorelines
spectacular roadcuts at 101/101A intersection
NH 9

grained aplite, both varieties of granite. The dikes are especially numerous near New Hampshire series plutons and probably originate by intrusion of magma from those bodies into country rock fractures.

Between Interstate 89 and Hillsboro (13 miles), you follow the Contoocook River upstream through broad, flat sections of valley deeply filled with sediments—glacial drift, deltaic and lake deposits of glacial Lake Contoocook. The original sediments were much thicker; much has been removed by downcutting of the river occasioned by postglacial rebound, as in the case of the Connecticut River. Hillsboro is the birthplace of Franklin Pierce, the 14th President of the United States, for whom New Hampshire 9 is named. Pierce was president from 1853-1857, when our flag had 31 stars, and our population was only 29 million.

Cuts between Hillsboro and South Stoddard (14 miles) are in the Kinsman quartz monzonite of the large Cardigan pluton of the New Hampshire series. This is one of the most striking igneous rocks of New England. Here it typically contains rectangular white potash feldspar crystals up to four inches long, suspended in a finer-grained, darker matrix rich in biotite mica with large red garnets. The strong alignment of the elongate feldspar crystals apparently results from laminar flow of the magma before its complete consolidation. Near contacts, the igneous rock commonly contains xenoliths of all sizes and shapes, pieces of country rock that were engulfed by the magma and then frozen in suspension.

Between South Stoddard and Keene (15 miles), you cross another broad swath of Littleton formation. This is one of the most widely distributed units of New Hampshire, nearly half of its bedrock floor from the Presidential range to the Massachusetts border.

Keene lies in the broadest section of the Ashuelot River valley, the site of yet another glacial meltwater lake called Lake Ashuelot. Thick sedimentary deposits underlie the valley floor and lakeshore features and terraces mark the valley sides. Varved clays that record at least 200 years of accumulation have been quarried in the past for brickmaking.

Between Keene and Brattleboro (15 miles), the route crosses the exotic terrain of the Bronson Hill anticline, certainly the most important element in the plate tectonics interpretation of this region. The anticline continues southward into Massachusetts and Connecticut. In New Hampshire, it occupies a zone immediately east of the Connecticut River between Massachusetts and Lancaster, where it continues northeast across the state and into Maine. Within the state it is about 160 miles long by 15 miles wide. It is believed to consist of materials from a volcanic island arc that formed in early Taconian time and collided with the continent at the height of that same mountain-building event. Much later, it was caught in the vice between colliding ancestral North America and Europe as the proto-Atlantic Ocean closed during the Acadian event. Collapse of the volcanic archipelago against the eastern margin of the continent provided the compressive forces for the gigantic thrust faults that dominate western Vermont.

Principal rocks of the Bronson Hill anticline are greenschists—metamorphosed basalts—of the Ammonoosuc formation and the closely related metamorphosed granites of the Oliverian domes. In map view, the domes suggest a string of sausages among the Ammonoosuc and other metamorphosed volcanic and sedimentary rocks of Ordovician age. Also included are several New Hampshire series plutons, the largest of which, once again of Kinsman quartz monzonite, lies just south of this section of New Hampshire 9. The Littleton formation is also present in patches, but it rests on the older rocks and is preserved mainly where it has been dropped to a low level by faulting. As might be expected, the Bronson Hill anticline is tightly deformed and riddled with faults, nearly all of which trend more or less from north to south, parallel to the fold axis. The conspicuous north to south topographic grain along this stretch of highway is a modest erosional expression of this underlying bedrock structure. At least some ridges south of the road near the Connecticut Valley, however, appear to be glacial drumlins.

Spaulding Turnpike and New Hampshire 16: Portsmouth—Conway

77 mi. / 123 km.

Because of its peculiar geological setting, Portsmouth boasts one of the finest natural harbors in New England. The rugged coast is solidly supported by the extremely tough and erosion-resistant metamorphosed volcanic and sedimentary rocks of the Rye formation, granite that intrudes the Rye formation, and quartzite of the Kittery formation. The geologic age of these rocks is uncertain, but is probably Ordovician or Silurian. Rocks in Vermont and New Hampshire are aged by their pre-metamorphic origin, mostly by fossils. Portsmouth harbor lies in the drowned mouth of the Piscataqua River, which marks part of the boundary between New Hampshire and Maine. Most of the harbors on the United States northeast coast are drowned river mouths. This is a submerged coastline, on which sea level has been slowly rising since the Ice Age. Drowning here reaches 15 miles up the Piscataqua and also far up into its tributary Squamscott River forming a fine system of inland bays. Spaulding Turnpike crosses Little Bay three miles northwest of the city where the Bellamy, Cocheco, and Salmon Falls rivers join the Piscataqua River.

Most of the route between the bridge at Little Bay and Rochester (18 miles) is on folded slates, phyllites, and schists of the Merrimack group, of which the Rye formation is part; but they are poorly exposed. A few cuts south of Rochester are in strongly layered gray gneiss of the Fitchburg pluton, an enormous, northeast-trending granitic mass approximately five to ten miles wide that straddles the state from border to border and continues southward in Massachusetts giving it a total length of about 85 miles. The age of the pluton is uncertain, but it is thought to be part of the Devonian New Hampshire plutonic series.

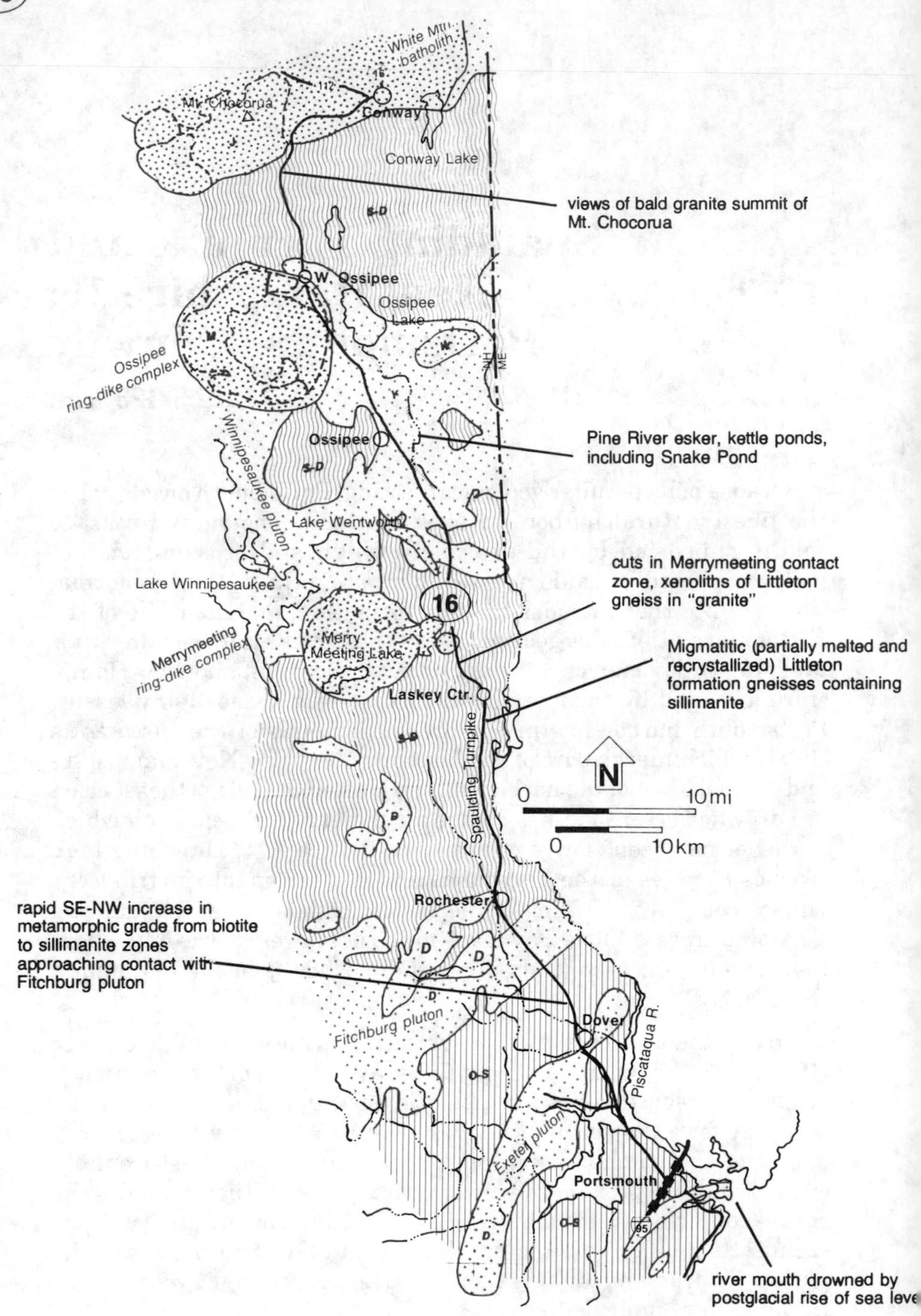

Spaulding Turnpike and NH 16
Portsmouth — Conway

Between Rochester and Laskey Center (11 miles), the Spaulding Turnpike crosses strongly metamorphosed and deformed Littleton schists. This same unit, with variations in rock type and grade of metamorphism, forms the bedrock foundation of fully half of southern New Hampshire and also holds up the summits of the Presidential Range. Numerous roadcuts expose a rather coarse-grained, strongly layered mica schist in which the mica flakes sparkle in the sunlight. Red garnet and needles of sillimanite are common mineral constituents, indicating extreme metamorphic temperatures. Large cuts 2 to 2½ miles north of Rochester exit 16 show considerable folding and faulting of the schist layers. Several rusty-weathered basalt dikes cut through the schist, one of them 8 feet wide. Dike rocks like these are extremely numerous in the state. They formed as fractures filled with magma from the White Mountain plutonic and volcanic complex. Not all the dikes are basaltic. Composition, color, and grain size of the dikes all vary widely, even including white aplites, fine-grained rocks of granitic composition. All formed at rather shallow depth where the host rocks tend to have open fractures that permit injection of rock melt. Less common are sills where magma forced its way between layers of the host rock. Pegmatites, very coarse igneous rocks, are also common in dikes throughout the state; but most are related to the earlier New Hampshire plutonic series.

Some of the roadcuts in Littleton schist near the north end of the turnpike are migmatitic, meaning metamorphic rocks that contain many thin interlayers and lenses of granitic rock. Most migmatites of wide areal extent are products of partial melting that occurs during high temperature metamorphism and produces localized pockets of magma.

Two miles north of the end of the turnpike, New Hampshire 16 just nips the eastern end of the Merrymeeting ring-dike complex of the White Mountain magma series. Large roadcuts reveal light gray granite with some "floating" inclusions, xenoliths, of darker gneiss that were torn from the intruded rock and engulfed in the magma. The main Merrymeeting body lies two miles farther west.

Between the New Hampshire 109 junction to Sanbornville and Ossipee (12 miles) are roadcuts in light gray Winnepesaukee quartz diorite of the New Hampshire plutonic series. The rock differs from granite in containing less potash feldspar and tending to be darker in color. Here it contains many inclusions of dark gray gneiss.

The area around Ossipee is famous for the Pine River esker that lies one to three miles east of the road. An esker is like an inverted river of sand. This one forms a continuous narrow ridge for about five

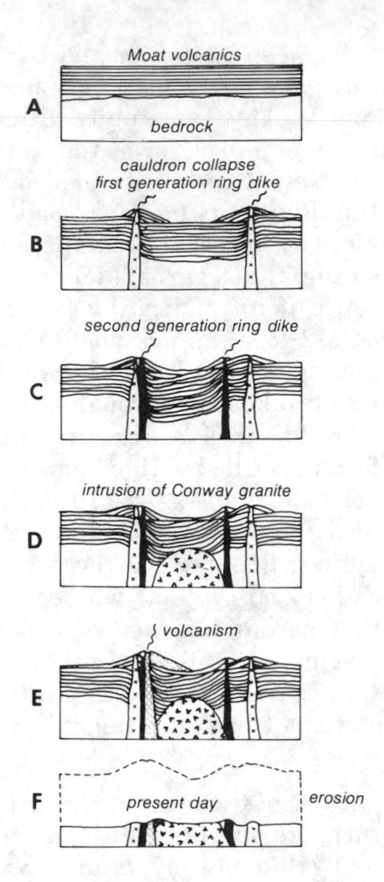

Schematic sequence showing the development of a ring-dike complex, characteristic of White Mountain magma series. There are many variations of the theme. Not shown is the deep-seated source of magma.
—From A. Quinn.

miles alongside the river. An erosionally isolated patch farther north near Ossipee Lake would make its full original length at least nine miles. Eskers form near the edges of wasting ice sheets through deposition of sand and gravel in the beds of meltwater streams that tunnel along the bottom of the glacier. When the glacier melts, these deposits are left as a ridge. The Pine River esker is visible from tiny Snake Pond, one mile east of New Hampshire 16, on New Hampshire 171, the road from Ossipee to Granite.

Many of the small lakes and ponds that dot this region are probably kettle lakes, formed where blocks of ice were left behind by the receding glacier. Such blocks often get buried in outwash sediments; when they finally melt, pits remain to mark their places.

The Ossipee Mountains loom large on the west side of the highway between Ossipee (junction New Hampshire 28) and West Ossipee (11 miles). The mountains are carved from the Ossipee ring-dike complex, another member of the White Mountain series that is almost perfectly circular in outline and approximately nine miles in diameter.

The origin of the White Mountain ring-dike complexes is one of the most interesting geologic stories of New Hampshire. Many of the igneous bodies, like the Wolfeboro and Ossipee plutons, are circular or oval in outline with Conway granite at their cores. Several also contain erosional remnants of Moat volcanic rocks. Many have one or more concentric ring-dikes of different composition. Hundreds of tabular dikes of variable composition cut the country rocks of the region around the bodies.

The most widely accepted theory of origin for these peculiar complexes has the Moat volcanic rocks erupting first from numerous vents, and eventually covering much of central New Hampshire with layered deposits thousands of feet thick. Eruption of volcanics from the magma chamber caused more or less circular blocks of the upper crust to sink into the magma chambers. Then magma welled up around these plugs to form the ring dikes, as well as a new series of eruptions. In some cases, two or more such roof collapses resulted in the formation of additonal ring-dikes.

In general, the earliest dikes are composed of black basaltic rock, the later ones of pale rocks that approach granite in composition. The tabular dikes formed by intrusion of the same magmas into straight fractures in the surrounding rocks. The Conway granite magma, meanwhile, was slowly rising into the plugs, probably making room for itself by magmatic stoping, a process in which broken blocks of the chamber roof sink while the magma rises to fill the void. The magma reached a level of about 4 miles below the surface before becoming completely solid, invading even the Moat volcanics in places.

Radiometric dating of the rocks shows that the White Mountain series rocks were emplaced between 185 and 110 million years ago, in late Triassic to Cretaceous time. Erosion since their emplacement has drastically reduced the landscape. All of the thick blanket of Moat volcanic rocks is gone except in the plugs that dropped the farthest, where the erosional level has not yet reached the base of the volcanic pile.

The Ossipee complex is one of the best examples of a ring-dike assemblage. The core rocks include the Conway granite, Moat volcanic rocks, and Winnepesaukee granite which contain remnants of older country rock. The core is completely circled by a narrow ring-dike of Albany quartz syenite, a rock of light greenish color when fresh, with relatively large rectangular crystals of potash feldspar suspended in a fine-grained groundmass. Except for narrow zones of Littleton schist on the north and south sides, the Winepesaukee granite entirely encloses the complex.

Between West Ossipee and Conway (14 miles), the bald granite peaks of the Chocurua Mountains at the east end of the Sandwich Range are open to view. Because of their accessibility and the good visibility from their bare summits, these mountains are among the most delightful climbing areas in New Hampshire. They are carved from the Mount Osceola granite of the White Mountain batholith. In map view, the batholith appears to be a cluster of numerous, large and small ring-dike complexes. Its irregular borders enclose a host of different granitic rocks, large remnants of Moat volcanic rocks, and large patches of older rock, including the Kinsman quartz monzonite and Littleton schists.

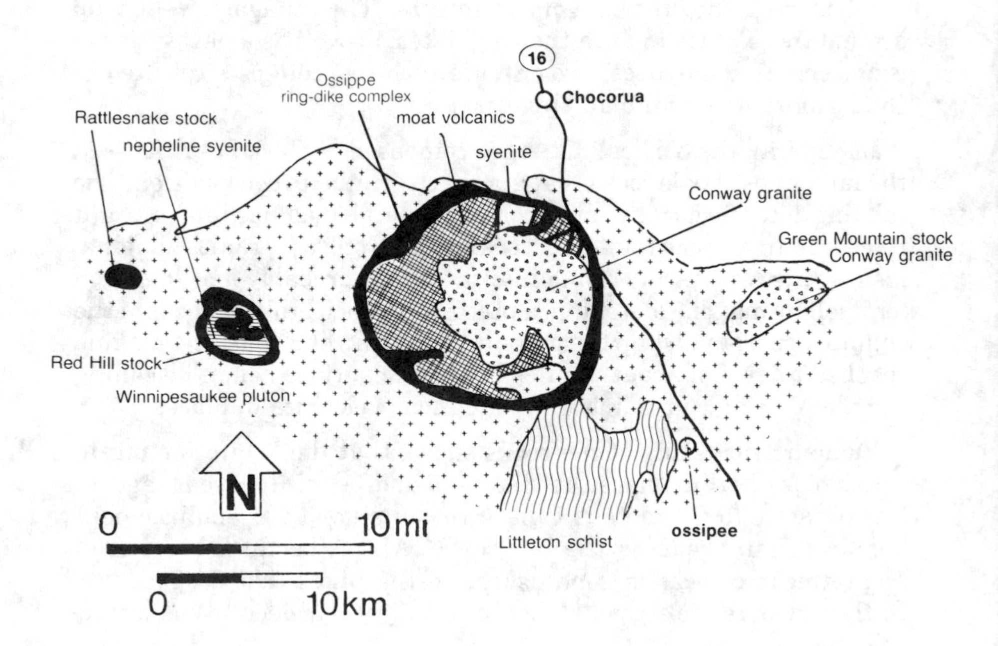

New Hampshire 16: Conway, N. H.— Maine Border North

69 mi. / 110 km.

Mount Washington Valley and Pinkham Notch Highway

Between Conway and Glen (11 miles), New Hampshire 16 follows the beautiful Mount Washington Valley of the Saco River upstream. The Saco Valley floor here is flat and very broad because it is filled to considerable depth with glacial outwash and glacial lake sediments deposited during recession of Wisconsin ice. The deposits were originally much thicker. Much of the sediment has been removed by down-cutting of the river during postglacial rebound of the land. Striking river terraces on the valley sides near North Conway show the original height of the fill.

This section of highway lies entirely within the Conway granite of the White Mountain batholith, the largest body of the White Mountain magma series. A batholith is simply a very large, irregular, igneous intrusive body; smaller ones are called stocks. This one appears to be made up of a cluster of numerous ring-dike complexes similar to the Ossipee and Wolfboro ring-dike complexes.

The large mountain mass four miles west of North Conway is Moat Mountain, the site of the thickest known deposits of Moat volcanic rocks. Other large exposures of these rocks exist in the Ossipee Mountains, 16 miles to the south, and on Pequawket Mountain, seven miles northeast of Moat Mountain.

Smaller bodies of Moat volcanic rocks are in the Belknap Mountains on the south side of Lake Winnipesaukee, on Mount Hale four miles southeast of Twin Mountain, and three miles southeast of Mount Lafayette. All of these isolated masses are in the dropped cores of ring-dike complexes, where they are sheltered from erosion. All are

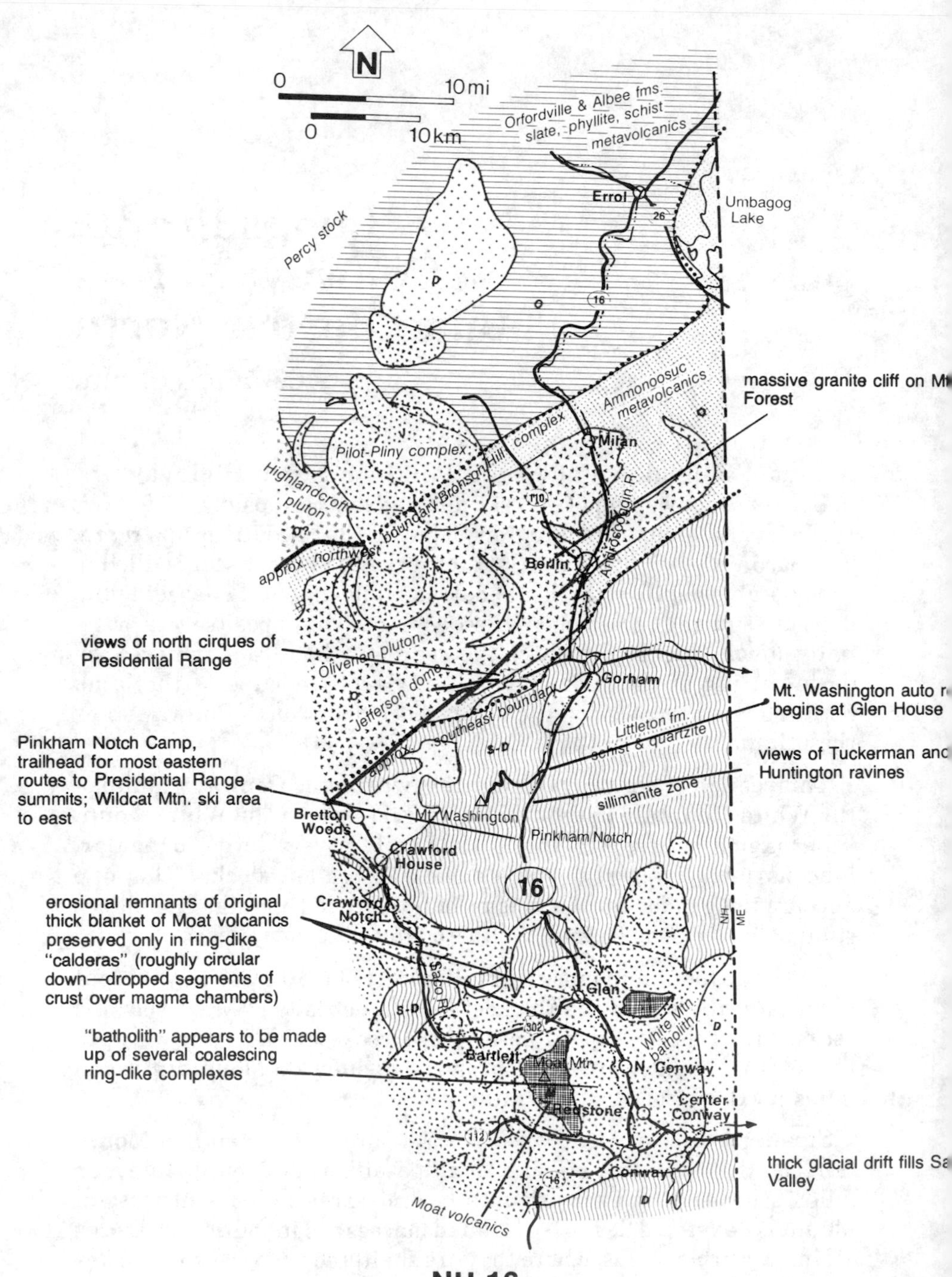

NH 16
Conway — Maine Border

considered to be remnants of a once continuous blanket of horizontally stratified volcanic rocks thousands of feet thick that covered most of central New Hampshire. They are the only surviving representatives of volcanic outpourings that signaled the beginning of White Mountain magmatic activity. The original thickness of the volcanic blanket is unknown, but it probably was at least 9000 feet, the estimated depth of the Moat Mountain deposits. Stratification in the various exposures tilts because each ring-dike plug collapsed unevenly. The rocks are variable, including many light-colored rhyolites, the volcanic equivalent of granite, and darker basalts and andesites which are largely confined to the Ossipee Mountains. They include both lava flows, resulting from quiet eruptions, and fragmental deposits from explosive vulcanism.

Moat volcanic rocks are approximately the same age as the lavas of the Connecticut Valley and record the same plate tectonic event—the separation of eastern North America from Africa that began in late Triassic time, about 200 million years ago. The lavas ascended along faults opened by stretching of the crust as the joined continents began to split.

The 22-mile section of New Hampshire 16 between Glen and Gorham is known as the Pinkham Notch Highway. Between Glen and Jackson (2 miles), the road continues over Conway granite of the White Mountain batholith.

Between Jackson and Pinkham Notch Camp (10 miles), all of the bedrock is Littleton schist and quartzite, the same tough rocks that hold up the Presidential Range. There are only a few roadcuts. Numerous trails lead to the various routes up Mount Washington; one, on the east side of the highway, leads to Wildcat Mountain (4397

Faulted sillimanite-bearing Littleton schist near the north end of the Spaulding Turnpike. Pale granite intrudes the schist near the base of the cut.

Basalt dike in Littleton schist near north end of Spaulding Turnpike.

feet). Several shorter walking trails lead to such scenic spots as Lost Pond and Glen Ellis Falls. Pinkham Notch is a north-south-trending valley carved by the same kind of stream and glacial action that produced Crawford Notch on the other side of the Presidential Range. It is, however, about 700 feet higher and has a cross sectional profile much farther from the ideal shape of a glacially-gouged stream valley. There are several possible reasons for the differences. The valley, first of all, is less in line with the southeasterly advance of the Wisconsin ice sheet. Secondly, the ice sheet had to overtop the Presidential Range to get here, and then it must have moved in a more eastward direction down the mountain slopes, athwart the Pinkham Notch trend. Thirdly, the Littleton schists, gneisses, and quartzites here are apparently tougher and more resistant to erosion than the Conway granite of Crawford Notch. Finally, the broad, gently-sloping basin around Bretton Woods and funnel-like "Gate of the Notch" undoubtedly served to crowd ice into Crawford Notch and made the ice tongue a more efficient gouger than that of Pinkham Notch.

The best highway views of Mount Washington's Tuckerman Ravine are between Pinkham Notch Camp and Glen House (3 miles). This is the most famous of the cirques, or rock amphitheaters of the east side of the Presidential Range. Small glaciers that followed the existing stream valley carved it during early, and possibly also late, Wisconsin time. Other cirques exist on the north and south flanks of the range, but none on the west side. The best possible view of the range is from the summit station of the Wildcat ski area gondola. Mount Washington protrudes like a pyramid from a rather flat surface called the Presidential upland, interpreted as a preglacial erosional surface. You can see Tuckerman and Huntington ravines in true perspective, as though they were carved by swipes from an enormous ice cream scoop into the edge of the upland.

From the highway, the Presidential Range is especially open to view around Glen House, at the base of the Mount Washington Auto Road.

Between Glen House and Gorham (8 miles), the road follows the Peabody River downstream through a valley much wider and more open than that of the Ellis River south of Pinkham Notch, the drainage divide. The Peabody River joins the Androscoggin River at Gorham. The southern half of this section of road is over Littleton schists, and the northern half is on granite of the New Hampshire series plutonic rocks that intrude the schist. Roadcuts are scarce.

The highway follows the Androscoggin River between Gorham and Errol (36 miles). Its headwaters are in Umbagog Lake. At Berlin, seven miles from Gorham, evenly-spaced rock piles in midstream are not natural deposits. Mainly because of the availability of cheap water power this has been a paper mill town since the early 1800s. In the early days before trucks, two competing companies floated logs to the mills. A log boom anchored to the rock piles divided the river into two "avenues," one for each company.

Berlin is on the eastern flank of Jefferson dome, the largest of the Ordovician Oliverian granite bodies. Numerous intrusive bodies ar-

Glen Ellis Falls in sillimanite-bearing Littleton schist near Pinkham Notch.

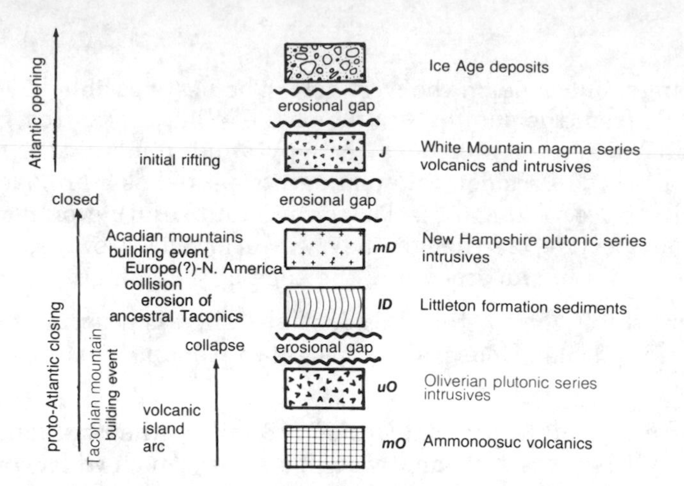

Time sequence for formation of major rock units of central New Hampshire and their relationship to plate tectonic events.

ranged like a string of sausages follow the axis of the Bronson Hill anticline. Near the village, the granite is interleaved with metamorphosed volcanic rocks, the Ammonoosuc formation. They are part of a volcanic island arc that formed in the early phases of Taconian mountain-building and was later shoved against and welded to what was then the continental margin.

The striking cliff on Mount Forest at the southwest edge of the village is granite of Jefferson dome. The route continues over this bedrock and Ammonoosuc formation to Milan (8 miles) and beyond. A few roadcuts north of Milan expose gray, banded gneiss; one contains a 5-foot black basaltic dike related to the White Mountain magma series. Three miles north of Milan is the boundary between metamorphosed volcanic rocks of the Ammonoosuc formation and metamorphosed sedimentary rocks, mainly schists, that belong to the Ordovician Albee formation. The latter rocks underlie the road the rest of the way to the Maine border (27 miles). There are almost no roadcuts. The route is mostly woodsy, but occasional open sections reveal flat wetland or hummocky fields of glacial till.

Tuckerman Ravine from NH 16 near Glen House.

Mount Washington Auto Road

16 mi. / 26 km. round trip

For the roadside geologist in New Hampshire, a trip to the summit of Mount Washington along the 8-mile long auto road is a must. The company that maintains the road charges a toll. The road is mostly gravel with a few stretches of pavement, and it climbs at a steady 12% grade for approximately 4600 vertical feet. Another way up is by the Mount Washington cog railway, which is even steeper. Timberline is at 4500 feet, and the beautiful scenery above that level is open and far more alpine than one might expect in the eastern United States. On especially clear days it is possible to look into Maine, Vermont, New York, Massachusetts, and Canada, and to see the sun reflecting from the Atlantic Ocean.

The alpine character of Mount Washington and the Presidential Range it dominates is largely due to its weather. The summit lies in the path of the principal storm tracks and air mass routes that affect the northeast, and the great mass of the mountains traps the weather. The climate above timberline is subarctic, ecologically similar to that of northern Labrador. The Mount Washington Observatory has recorded the most severe combinations of wind, cold, icing, and storminess ever reported in the world. The wind there exceeds hurricane force of 75 mph on an average of 104 days each year. The world record wind velocity of 231 mph was registered there on April 12, 1934.

Bedrock along the entire auto road is early Devonian Littleton formation, interbedded mica schist and quartzite, formed through severe metamorphism of shales and sandstones. These are among the most compact and resistant rocks of New England. They support all of the high summits of the Presidential Range, as well as a number of

Mt. Washington Auto Road

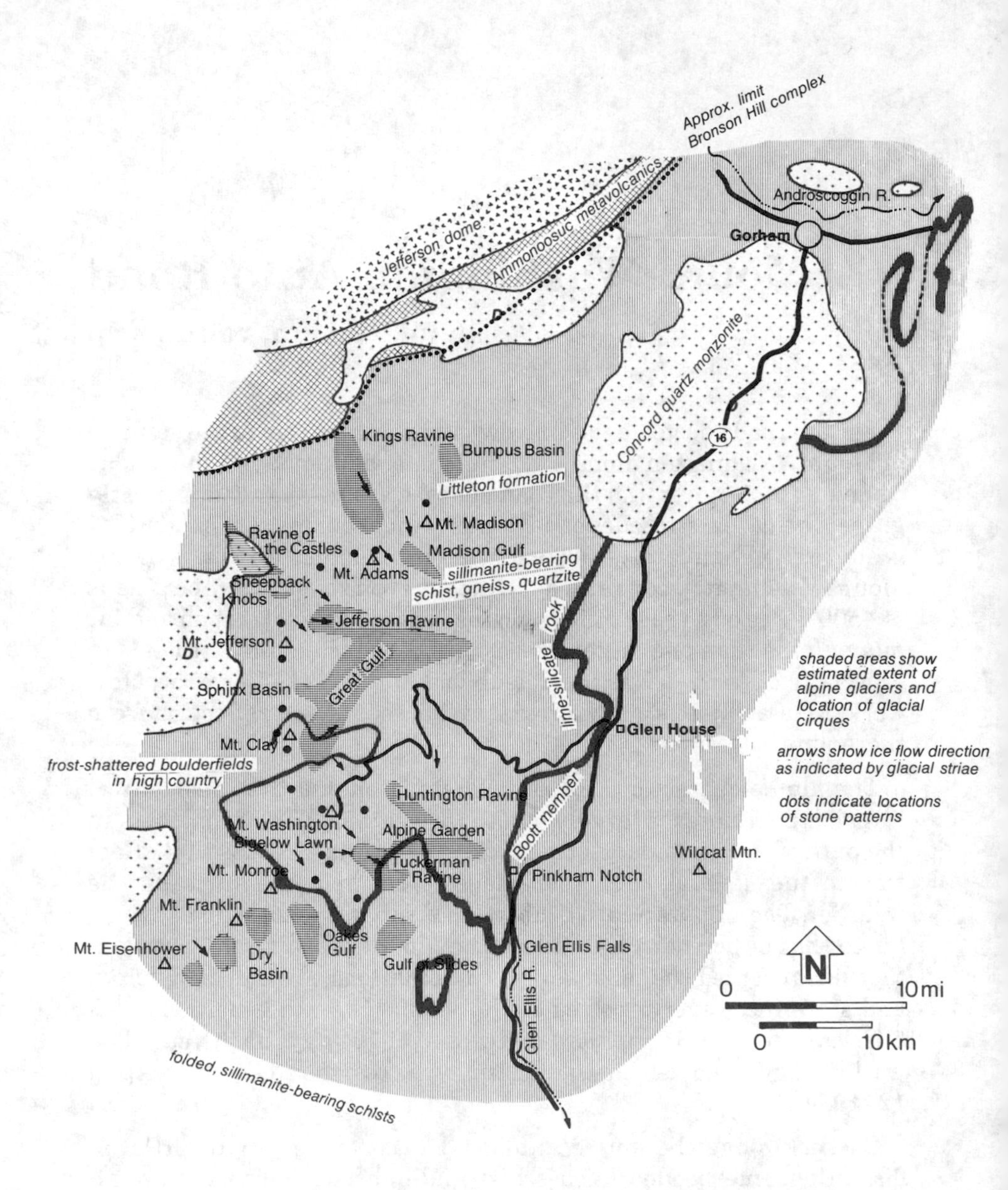

lesser peaks outside of the range, including the famous Mount Monadnock of southern New Hampshire.

The schists locally contain the minerals andalusite or sillimanite in abundance, both indicators of very high temperature metamorphism. Geologists estimate that metamorphism occurred at a depth of seven miles and a temperature of about 1000 degrees Fahrenheit (540 degrees Centigrade) during Acadian mountain-building approximately 380 million years ago. These rocks were in the core of the rising Acadian Mountains. Erosion of the overburden occurred over the next 200 million years while much of the resulting sediment spread westward displacing a shallow inland sea and forming the Catskill delta, now best preserved in New York and Pennsylvania. Were the mountains seven miles high, you might ask? Probably not. Most mountains are made of low density rocks and have deep roots projecting into the denser rocks of the lower part of the Earth's crust. They float like colossal icebergs! And, like icebergs, they float higher as erosion shaves their tops. So little by little the Acadian Mountains kept floating up as erosion destroyed them. Spread over 200 million years, the rate of erosion would average 1.9 feet every 10,000 years. As a comparison, one estimated current rate of erosion of the Colorado Plateau in the Grand Canyon country of Arizona is more than 5 feet per 10,000 years.

Numerous boulders or outcrops along the auto road contain the critical mineral sillimanite, the basis of much of this interpretation. In some deeply weathered exposures above timberline, the whitish, elongate crystals are etched in relief and clearly visible. Most are partially, or wholly, replaced by fine-grained white mica and quartz,

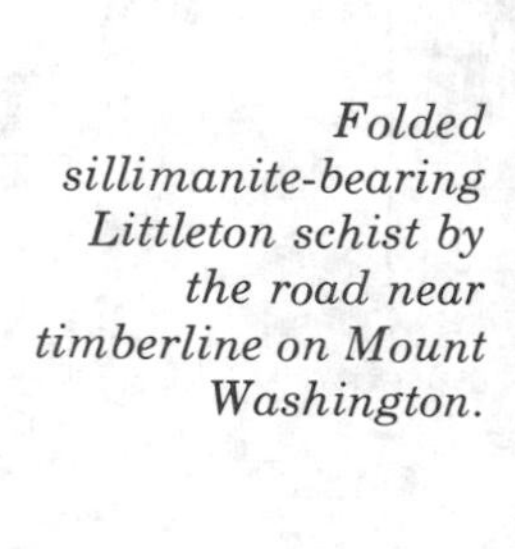

Folded sillimanite-bearing Littleton schist by the road near timberline on Mount Washington.

although the original sillimanite crystal form is preserved. This common replacement happens when the minerals become unstable at lower metamorphic temperatures and are replaced by stable minerals.

Also visible along the auto road and in many places above timberline in the Presidential Range is the extensive and well-defined folding of the Littleton formation, particularly in some outcrops between the 5 and 6 mileposts. In crumpled sillimanite-bearing mica schists, the sillimanite crystals are commonly bent around the small folds, indicating that some deformation outlasted their crystallization. Nevertheless, the strong parallel alignment of the sillimanite crystals suggests that most of them were oriented by the stresses of deformation, that they grew while the rock was being folded.

The most striking geologic features of the Presidential Range are glacial. Tuckerman Ravine, Huntington Ravine, Great Gulf, Sphinx Basin, Jefferson Ravine, Madison Gulf, and the Gulf of Slides are all rock amphitheaters, or cirques, along the east side of the range; most are visible from the road or summit. Several other similar basins lie

Maps of different stone patterns relative to slope angles. (A) NETS form on slopes of less than 3 degrees, this one from Bigelow Lawn one mile south of Mount Washington summit at an altitude of about 5460 feet. (B) STRIPES form at angles of 3-7 degrees, Bigelow Lawn at about 5400 feet. (C) LOBES form at angles of greater than 7 degrees, northwest of Mount John Quincy Adams at about 5300 feet. —From R.P. Goldthwait.

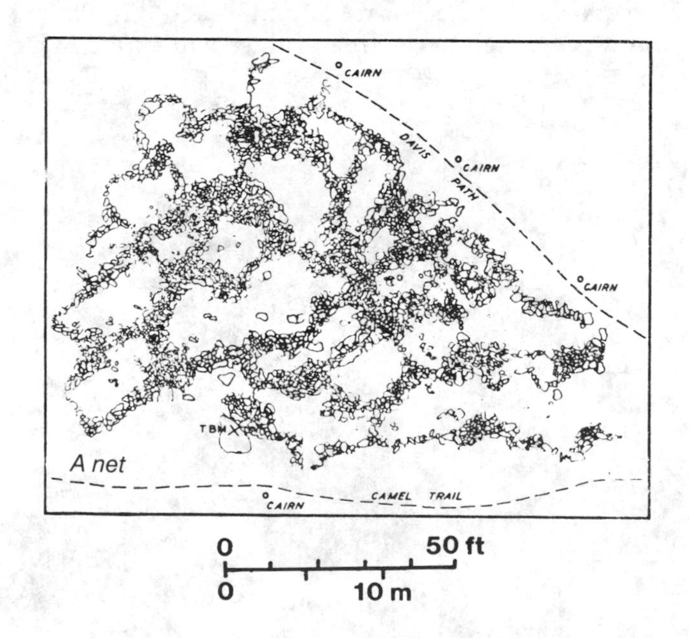

at the northern and southern ends of the range. All were carved by small alpine glaciers that coursed down the existing stream valleys during Wisconsin glaciation. The same thing happened on other high mountains of the northeast, notably Mount Katahdin of Maine and the high peaks in the Adirondacks of New York. At the height of the Wisconsin glacial stage, the mountains, in fact the whole New England region, were buried under a massive ice sheet that reached southward to Long Island. The evidence of that exists in glacial striations and grooves, stream-lined rock knobs called "sheepbacks," rounded ledges, and erratic boulders carried in the ice and left near the mountaintops when it melted. Small lakes occupy basins scooped out of the high mountains by the ice; Lakes of the Clouds southwest of

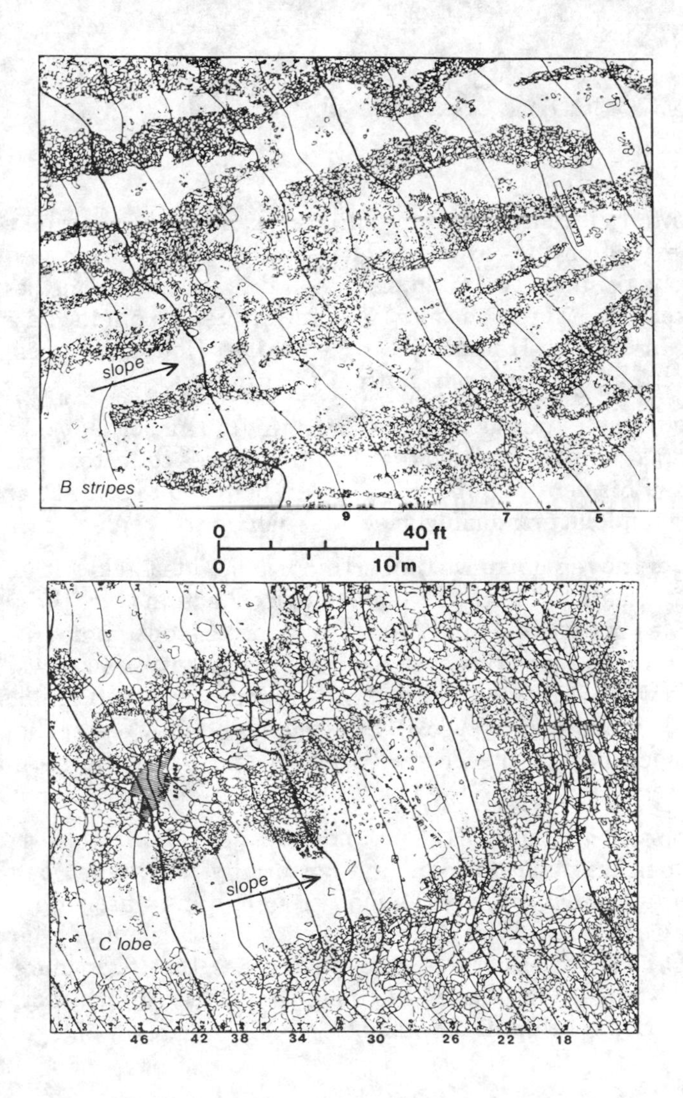

View northwest from near timberline on Mount Washington auto road across the Great Gulf to Jefferson ravine. Mount Jefferson on left; Mount Adams rises to right out of view.

the summit of Mount Washington are examples of such lakes. The best views of cirques are of Jefferson Ravine and Great Gulf from numerous points between timberline and the summit. The best view of Tuckerman Ravine, the most famous cirque, is from the base of the mountain on New Hampshire 16 near Glen House. It appears as a very impressive deep gash south of the summit.

The summit pyramids of Mount Washington and other peaks in the Presidential Range jut from rather flat surfaces collectively referred to as the Presidential upland. These are believed to be remnants of a once continuous erosional surface developed in preglacial time.

Most of the region above timberline is a boulder field with little or no intact bedrock. This is largely due to frost action since the Ice Age, which also obliterated much of the evidence of continental deglaciation. As anyone who has ever had frozen water pipes knows, water expands when it freezes. The summit region here is subject to a lot of freezing and thawing, especially in spring and fall. Water that penetrates openings in the bedrock freezes and pries the rock apart; countless repetitions in the last 12,000 years have shattered them.

Many stone patterns in the uplands are also caused by frost action. The stones are pushed around and gradually rearranged by the expansion and contraction of the matrix soil as the water in it freezes and thaws. The patterns on level or very gently sloping ground are polygonal. On slopes of three to seven degrees, they form more or less parallel stripes. On steeper inclines, they tend to form horseshoe-shaped lobes that point downslope.

New Hampshire 101: Hampton—Keene

90 mi. / 144 km.

also see map page 180

Hampton lies near the rocky New Hampshire shore at the southwest plunging tip of the Rye anticline. The rocks are erosion-resistant schists and gneisses of the Rye formation of probable Ordovician-Silurian age. The unit crops out along the shore and in a few small bedrock knobs that project through the salt marshes. The grade of metamorphism changes rapidly in this region, ranging from sillimanite zone north of Rye Beach through staurolite, garnet and biotite at Hampton. It is rare to find metamorphic temperatures varying so greatly in such a short distance.

Between Hampton and Manchester (38 miles), you cross the flat Seaboard section of the state and then gradually mount the lower edge of the New England uplands. Marine sediments and shoreline features in the Seaboard section show that the area was partially submerged under Atlantic waters near the close of the Ice Age.

Nearly all the bedrock between Hampton and Raymond (20 miles) is poorly exposed schist of the Merrimack group, again of probable Ordovician-Silurian age. The exception is dark diorite of the Exeter pluton, exposed in old roadcuts just west of Exeter (7 miles west of Hampton). Near Raymond, you cross compressed mineral zones again, going from chlorite to sillimanite in two to three miles. All of the aluminum-rich metasedimentary rocks farther west to Keene, derived principally from original clay muds, record metamorphism at temperatures high enough to form sillimanite.

Most of the route between Raymond and Ponemah (34 miles) follows the axis of the large Fitchburg pluton. This is an enormous granitic mass approximately 5-10 miles wide that straddles the southeastern corner of the state and continues southwest into Massachusetts, a total length of about 85 miles. The original igneous

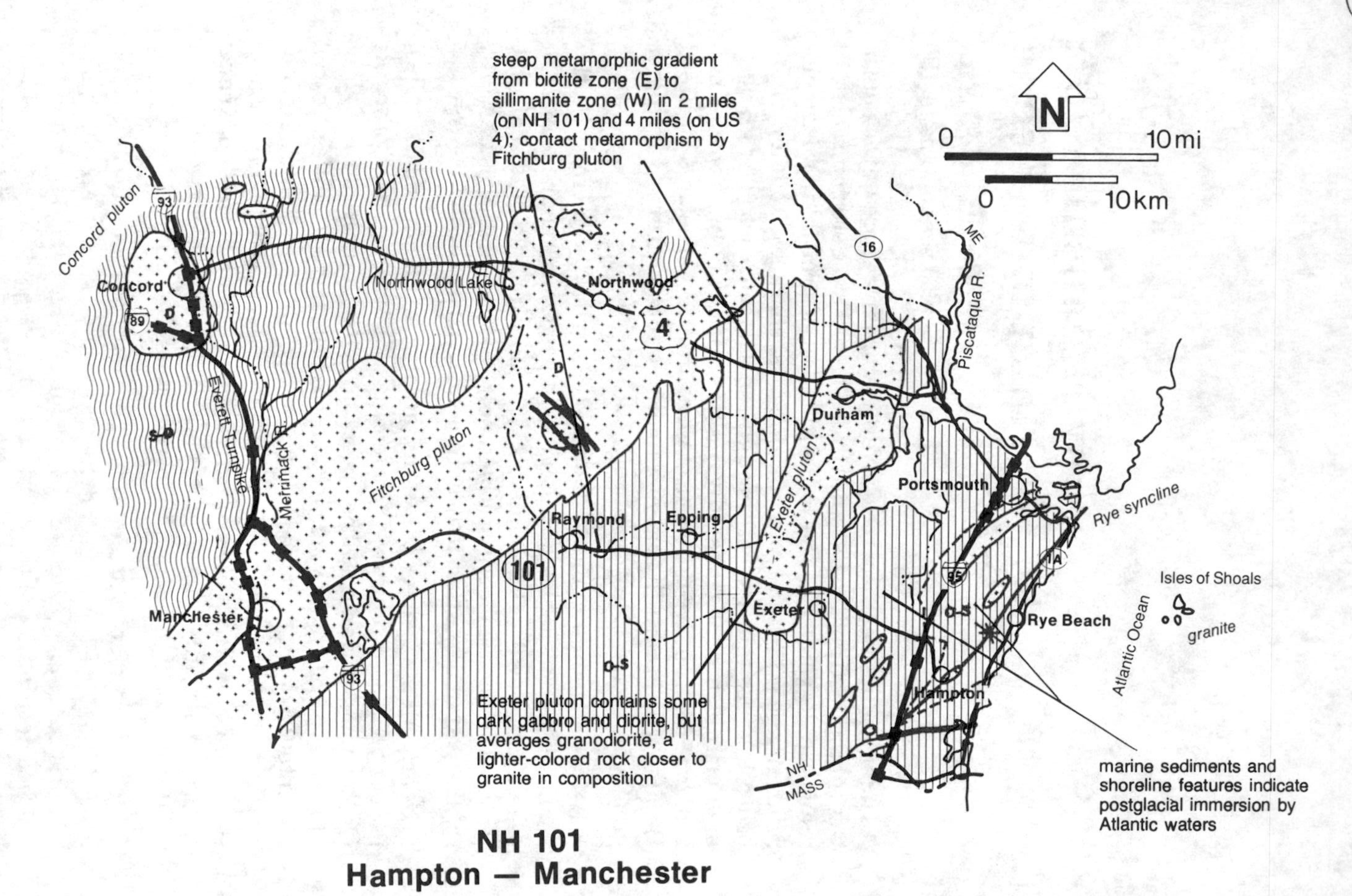

NH 101
Hampton — Manchester

Complexly deformed gneiss within Fitchburg pluton at intersection NH 101 / 101A.

intrusion is thought to be Devonian age and it is tentatively assigned to the New Hampshire plutonic series. Some characteristics of the Fitchburg rocks are well-illustrated in very large roadcuts near Ponemah, at the intersection with New Hampshire 101A. Here the rock is a complex array of grayish, strongly layered and locally banded gneiss. It contains large fragments of schist and is shot through with white granitic veins and dikes, many of which are pegmatites containing large books of mica. Locally, the rocks appear to have been intensely sheared and granulated and then healed by the metamorphism.

Manchester is in the Merrimack Valley, which has a glacial history much like that of the Connecticut Valley. During Wisconsin deglaciation, glacial Lake Merrimack extended 80 miles upstream from Lowell, Massachusetts, to Plymouth, New Hampshire, and seawater reached almost to Manchester. Thick sedimentary deposits underlie the flat floor, and several levels of terraces as well as shoreline features adorn the valley sides. The originally level water plain is now about 400 feet higher at Plymouth than at Manchester, a result of postglacial uplift. As in the Connecticut Valley, downcutting of the Merrimack River exposed numerous sections of varved clays that enable geologists to chart ice recession. For example, correlation of varves indicates that retreat of the ice margin from Concord to Franklin, a straight-line distance of 20 miles, took 500 years.

With only minor exceptions, all of the bedrock in the low, hilly country between Ponemah and Keene (39 miles) is Littleton schist with sillimanite, much of it gneissic and rather coarse-grained owing to the high-grade metamorphism. An excellent place to examine the bedrock and see the landscape for miles around is on top of Pack Monadnock Mountain (2288 feet) in Miller State Park just off the highway 17 miles west of Ponemah, about four miles east of Peterborough. The summit is easily reached by a steep, paved road for which there is a toll.

This mountain stands considerably higher than the surrounding countryside, as do its larger namesakes, Monadnock Mountain (3165 feet) 10 miles to the west and the lesser Little Monadnock Mountain (1883 feet). Many other similar knobby mountains jut conspicuously from the New England uplands, including even Jay Peak of northern Vermont. All are interpreted as residuals of resistant rock that survived a general downwasting of the landscape. The Littleton formation supports many such isolated mountains in southern New Hampshire. If you accept the idea that they mark resistant bodies of rocks, you have to assume that the Littleton formation varies greatly in its resistance to erosion. The same formation supports the high peaks of the Presidential Range where the rocks closely resemble those of Pack Monadnock.

Highly sheared gneiss at intersection intersection NH 101 / 101A.

Rocky Gorge carved by Swift River along joints in granite probably belonging to a Concord pluton, NH 112 west of Conway.

New Hampshire 112: Kancamagus Highway Conway—Lincoln and Lost River Road Lincoln—Junction US 302

57 mi. / 92 km.

The scenic Kancamagus Highway connects Conway on the east with Lincoln on the west (34 miles). The route lies entirely within the borders of the White Mountain batholith, except for the westernmost mile. The batholith is the largest intrusive body of the White Mountain magma series. In reality, it appears not to be a single body, but a composite of numerous large and small ring-dike complexes that ran interference with each other. The dominant rock unit within the batholith is the Conway granite, which underlies most of the Kancamagus Highway.

Between Conway and Passaconaway (west of the intersection with Bear Notch Road—13 miles), the route passes between Moat Moun-

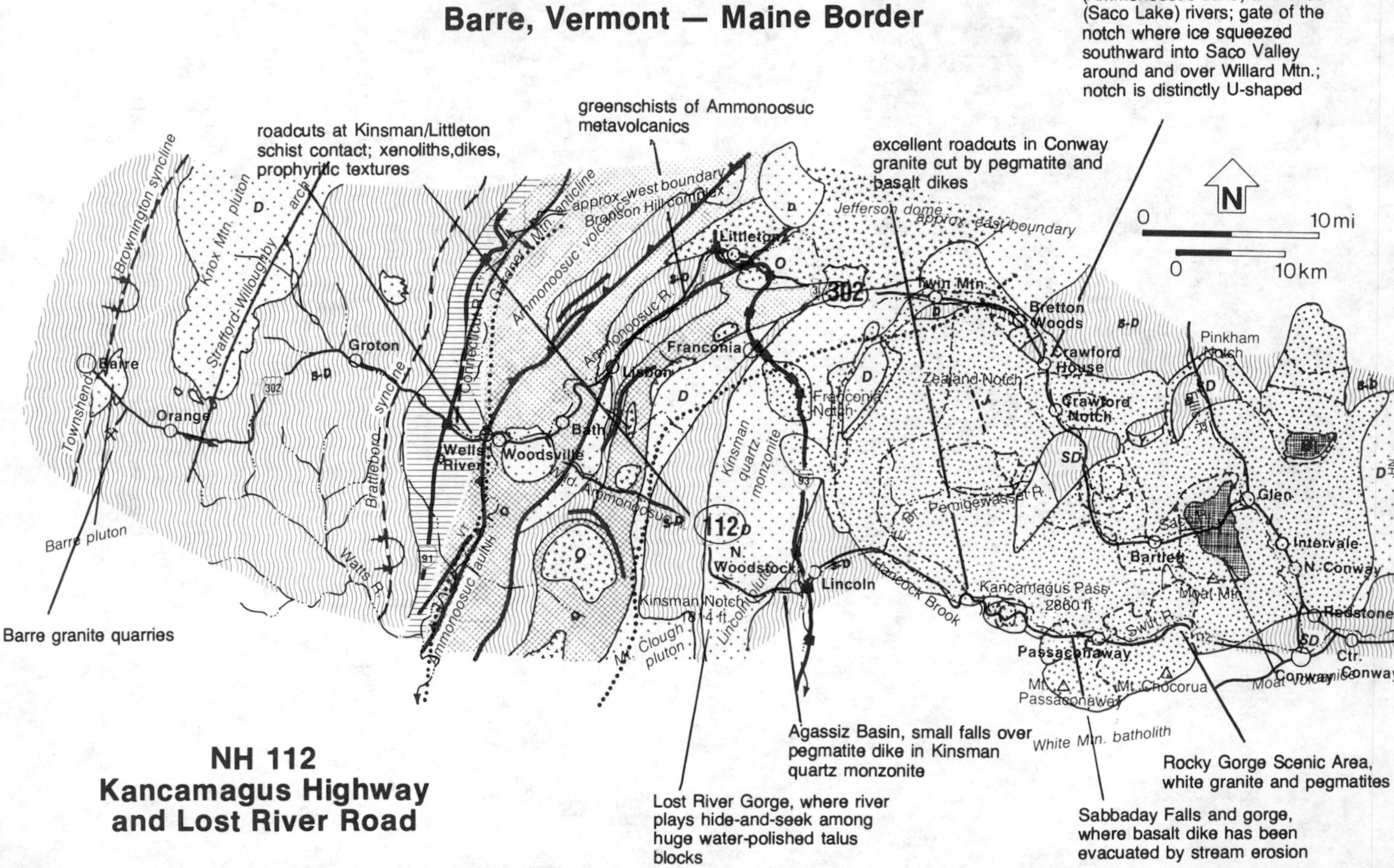

US 302
Barre, Vermont — Maine Border
NH 112
Kancamagus Highway
and Lost River Road
headwaters of Ammonoosuc (Ammonoosuc Lake) and Saco (Saco Lake) rivers; gate of the notch where ice squeezed southward into Saco Valley around and over Willard Mtn.; notch is distinctly U-shaped
greenschists of Ammonoosuc metavolcanics
roadcuts at Kinsman/Littleton schist contact; xenoliths,dikes, prophyritic textures
excellent roadcuts in Conway granite cut by pegmatite and basalt dikes
Barre granite quarries
Agassiz Basin, small falls over pegmatite dike in Kinsman quartz monzonite
Lost River Gorge, where river plays hide-and-seek among huge water-polished talus blocks
Rocky Gorge Scenic Area, white granite and pegmatites
Sabbaday Falls and gorge, where basalt dike has been evacuated by stream erosion
N
0
10 mi
10 km
Barre
Orange
Groton
Wells River
Woodsville
Bath
Lisbon
Littleton
Franconia
Twin Mtn.
Bretton Woods
Crawford House
Crawford Notch
Pinkham Notch
Glen
Intervale
N. Conway
Redstone
Conway
Ctr. Conway
Bartlett
Passaconaway
Mt. Passaconaway
Mt. Chocorua
Kancamagus Pass 2860 ft.
Hancock Brook
Lincoln
N. Woodstock
Kinsman Notch
Mt. Clough pluton
Barre pluton
Townshend
Browningon syncline
Knox Mtn. pluton
Stratford-Willoughby arch
Brattleboro syncline
Watts R.
Connecticut R.
Ammonoosuc R.
Ammonoosuc volcanics
Bronson Hill complex
approx. west boundary
Jefferson dome
approx. east boundary
Zealand Notch
Franconia Notch
Kinsman quartz monzonite
E. Br. Pemigewasset R.
Swift R.
Saco R.
Moat Mtn.
Moat volcanics
White Mtn. batholith
ME
NH
302
112
93
91
D
S-D
SD
O

tain on the north and Mount Chocorua, south. Moat Mountain is the type locality for the Moat volcanic rocks, the place for which they are named. The mountain is largely made up of these rocks which apparently once formed a continuous blanket thousands of feet thick over most of central New Hampshire as a result of a long period of vulcanism about 175 million years ago, at the beginning of White Mountain magmatic activity. All that remain are a few erosional remnants preserved in the cores of ring-dike complexes where they have been protected from erosion. To the south, 3475-foot Mount Chocorua presents its several bald summits carved from Mount Osceola granite, the second most abundant rock type of the batholith. Because of easy access and good viewing from the peaks, Mount Chocorua is popular among climbers and hikers. Only one trail leads from this highway—two miles east of Passaconaway, but there are seven trails on the other side of the mountain.

Rocky Gorge Scenic Area between Moat Mountain and Mount Chocorua has a parking lot and a delightful place to swim in the Swift River. The little gorge is carved along vertical joint fractures, probably in the Concord granite. Water-polished surfaces alongside the gorge show that the rock is rather fine-grained and white and contains a number of large masses of white pegmatite, very coarse rocks composed almost entirely of the minerals quartz and feldspar. Pegmatites are believed to form principally by injection of water-rich magma into fractures. The Concord granite here is part of a very small body within the batholith, but it belongs to the much older New Hampshire plutonic series.

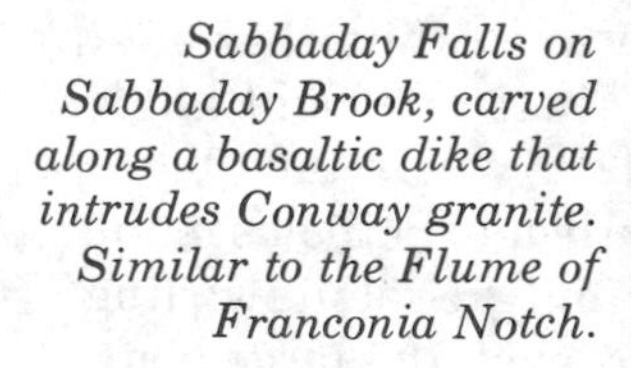

Sabbaday Falls on Sabbaday Brook, carved along a basaltic dike that intrudes Conway granite. Similar to the Flume of Franconia Notch.

Basalt dike in Lincoln pluton, 12 miles east of North Woodstock on NH 112.

The broad valley around Passaconaway, called Albany Intervale, affords some open mountain views, including some bare granite summits and glacially-shaped sheepback cliffs. The cliffs are nearly all south or southeast-facing, whereas the south side of the valley has relatively gentle slopes. The same asymmetry is visible on innumerable mountains of the northeastern United States. It is caused by progression of the Wisconsin ice sheet over the landscape scraping smooth the slopes that faced upstream while plucking rock fragments from the downstream sides, leaving cliffs. The term sheepback normally refers to much smaller rock knobs, but the asymmetry of whole mountains may be fashioned in the same way.

Many hiking trails lead both north and south from the highway between Passaconaway and Kancamagus Pass (9 miles). One of the most delightful short walks is to Sabbaday Falls, less than 2½ miles from Passaconaway. The falls and brook were so named by early explorers who discovered them on the Sabbath day as they sought a way across the mountains to the south. The falls have carved a deep, narrow gorge along basaltic dikes in the Conway granite that offer less resistance to erosion than the host rock. There is a right angle bend in the stream channel above the lower falls where the dike is offset, and the upper falls cascade over a granite ledge. Potholes in the streambed below the falls were carved by sand and gravel in swirling eddies, principally in times of flood. Incredibly, even the large boulders that clutter the channel downstream may be moved by torrential flooding.

The climb up 2660-foot Mount Potash, which begins along Downes Brook a little more than a mile east of the Sabbaday Falls trail, is a rewarding short trip. From several ledges en route, or from the almost bare summit, you can see 4060-foot Mount Passaconaway (south) with a landslide scar on its side, 2520 foot Hedgehog Mountain (east), 3200-foot Mount Paugus (southeast), and the tip of 3475-foot Mount Chocorua (east). An interesting "rocking stone" on the southeast side of the Potash summit, a huge glacial erratic of Conway granite, can be rocked by hand.

At Kancamagus Pass (2855 feet) is a parking and picnic area with a splendid view of the mountains to the north and northeast. Nearby, large roadcuts expose very coarse-grained flesh-colored Conway granite and pegmatite dikes.

Between the pass and Lincoln (12 miles), the road descends 1900 feet along Hancock Brook and the east branch Pemigewasset River. Two miles west of the pass is an overlook to Mounts Hitchcock (3648 feet) and Hancock (4403 feet), more or less north. A glacial cirque on Mount Hancock has bare rock headwalls and talus-covered slopes below. The roadcut across from the overlook exposes pink Conway granite with numerous steeply-inclined joint fractures, some of which are filled with thin basaltic dikes.

The Pemigewasset Valley around Lincoln and North Woodstock is broad and open with lovely mountain scenery. From here, the sheep-back character is apparent in several mountains profiled to the west, the slopes that faced into the ice flow are more gentle than the plucked slopes on the downflow side. The bedrock at Lincoln is Littleton schist at the western margin of the White Mountain batholith.

Between North Woodstock and Kinsman Notch (6 miles), the Lost River Road climbs about 900 feet. Agassiz basin is a point of geologic interest in the Lost River about a mile west of North Woodstock. It was named for Louis Agassiz, who was one of the first to recognize that much of the northern hemisphere was once covered by massive ice sheets. The basin is a section of stream channel cut into Kinsman quartz monzonite and a pegmatite dike that cuts through it. The dike is more resistant to erosion than the host rock, and thus forms a cliff and cascade. The bedrock is Kinsman quartz monzonite in a large intrusive igneous body called the Lincoln pluton of the New Hampshire plutonic series. It is well exposed at and near the notch, a generally coarse-grained rock, gray in color, with lots of flecks of black biotite mica that sparkle in the sun. The rock locally contains large, rectangular, white grains of potash feldspar suspended in the finer-grained matrix, a porphyritic texture. Numerous suspended fragments of dark gray schist and gneiss are locally clustered in great confusion. The inclusions are a trademark of the Kinsman quartz monzonite visible in almost any exposure. They come in all sizes and shapes and are pieces of the country rocks, mostly Littleton formation, that were torn off and engulfed by the intruding magma. Coarse, white pegmatites also appear in some of the exposures.

Lost River Gorge, just south of the pass, is one of the true scenic pleasures in New Hampshire. It is under the protection of the Society for the Protection of New Hampshire Forests and is operated as a tourist site. The society maintains trails, boardwalks, stairs, and ladders to give easy access to the gorge, and an excellent small museum with geological dioramas and exhibits. Lost River is so named because, in the gorge, it tumbles through labyrinthine tunnels among immense boulders, disappearing and reappearing unpredictably. Over most of the way it can be heard underfoot, but not seen. The boulders are covered with mosses and lichen and overhung with foliage, enhancing their sculpture with color and shadow and giving them an aura of mystery.

Lost River Gorge, according to Andrew H. McNair, is an unusual glacial valley that was eroded in the quartz monzonite first of all, by a stream working its way along a group of parallel, east-west-trending fractures that slope steeply southward. When glacial ice moved over the area, the notch was gouged deeper and wider and given a more rounded profile. The steep cliffs that now overlook the north side of the gorge are products of glacial plucking by the overriding ice. Later, during deglaciation, the ice lingered longer on the Ammonoosuc, or north side of the pass than on the Lost River side, so that torrents of meltwater had no escape but eastward through the gorge. Most of the water sculpture, including potholes and the smooth hollow faces of

Lost River Gorge at Kinsman Notch. Smooth, curved forms were carved by torrents of glacial meltwater rushing through huge talus blocks.

Profusion of Littleton schist xenoliths in Kinsman quartz monzonite north of Kinsman Notch near Beaver Pond.

boulders, was done during this period. When the ice melted far enough down the Ammonoosuc Valley to open an outlet in that direction, all that was left was the small stream that flows through the gorge today, tumbling among the huge boulders. The fine-grained sediment had by then been almost completely flushed out by the roaring cataracts of glacial water.

Most of the blocks that clutter the gorge probably fell from the north cliffs in the period of ice wastage. The climate was colder then and the rocks more subject to destruction by frost heave, especially in spring and fall. The boulders are unusually large because the fractures of the cliffs are widely spaced. The steep southward slope of the fractures facilitated their collapse into the gorge.

Several geologic features are visible at Beaver Pond, three-quarters of a mile north of Lost River gorge. Exposed at and near the dam are ledges of Kinsman quartz monzonite containing fragments of Littleton schist. A ledge of Littleton schist about 100 yards north of the dam by the highway shows glacial scratches that reveal the ice flow direction was 20 degrees east of south. Glacial ice invariably contains much rock debris; the scratches result when the ice-bound fragments scrape against bedrock. A gravel bank at the north end of the pond contains stratified, water-worn sand and gravel released from the ice as it melted, and transported, sorted, and deposited by

(112) the resulting streams.

An unusual roadcut about one mile north of the Notch is almost precisely on the boundary between the Littleton schist and the Kinsman quartz monzonite. The intrusive nature of the quartz monzonite is abundantly demonstrated in parts of the long cut, where the magma penetrated along countless schist layers. In places, the schist was forcibly pried apart; yet most of the slabs maintain their original orientation. Others have been carried along by the flow and disoriented. Inclusions like these, particularly of the Littleton schist, are a trademark of the Kinsman quartz monzonite; the largest is believed to be 1½ miles long! Here, inclusions are more numerous and crowded than in most places because the exposure is right at the contact between the two formations. Farther away from contacts they are still numerous, but generally more scattered. Many blocks have border textures gradational with the quartz monzonite, the result of partial assimilation of the schist into the magma. The texture of the quartz monzonite here is coarse-grained and only locally porphyritic, unlike places in the interior of the intrusion where the larger mineral grains are commonly three inches long. Near contacts, there normally is too little time for such large crystals to form before the magma becomes completely solid.

Between Kinsman Notch and the junction with US 302 near Bath (11 miles), the road follows the Wild Ammonoosuc River downstream to its junction with the main branch Ammonoosic. Many of the stream boulders were undoubtedly glacially transported and dumped here when the ice melted. In time, they will be carried farther downstream by floodwaters as they are also broken down and abraded—but it will be a very long time!

Large block of Littleton schist in Kinsman quartz monzonite north of Kinsman Notch near Beaver Pond.

Glossary

Alluvium — unconsolidated, stream-laid sediments

Amphibolite — dark metamorphic rock chiefly composed of hornblende and plagioclase fieldspar

Anticline — upfolded rock layers

Asthenosphere — the weak part of the Earth's mantle that supports the strong upper mantle and crust (lithosphere) and behaves plastically

Augen — a term used to describe eye-shaped lenses of minerals, as in augen gneiss

Basalt — dark, fine-grained volcanic rock

Base level — the level below which a stream cannot erode; typically sea level, the level of a lake or master stream, or of a resistant barrier

Basement rocks — normally very old rocks upon which younger sediments and volcanic materials have accumulated

Batholith — a great irregular mass of coarse-grained igneous intrusive rock with an exposed surface of at least 40 square miles

Bed — sedimentary layer with more or less distinct rock type and upper and lower boundaries

Bedrock — continuous solid rock either exposed at the surface or overlain by unconsolidated sediments, part of the Earth's crust

Block faulting — type of faulting that divides the crust into blocks of different elevation along a series of steeply-inclined up-down faults, generally with parallel trends

Brittle — material that breaks abruptly with little or no change in internal shape; opposite of ductile, and characteristic of most rocks deformed at or near the surface

Calcite — calcium carbonate mineral

Calc-silicate rock — metamorphic rock mainly composed of calcium silicate minerals such as hornblende, epidote, and diopside, often with calcite or dolomite

Carbonate rock — rock composed of carbonate minerals, especially calcite and dolomite in limestones, dolostones and marbles

Chatter marks, glacial — rows of small arcuate cracks in bedrock caused by "chattering" of ice-bound rocks against it as the glacier moves

Chill zone — fine-grained border zone in igneous rocks where the melt cooled quickly against the country rock

Cirque — half-bowl-shaped rock amphitheater formed at the head of a valley, or alpine, glacier—typically with steep headwalls

Cleavage, mineral — tendency of minerals to split easily along planes of weak chemical bonding in their crystal structure

Cleavage, rock — tendency of rock to split along parallel, closely-spaced planes, characteristic of slates

Coastal plane — a low plain of little relief next to the ocean and covered with sediments; its continuation offshore is the continental shelf

Composite stratigraphic column — graphical composite of two or more rock sections with rock units "stacked" in order of oldest at the bottom, progressively younger upwards

Conglomerate — coarse-grained sedimentary rock, lithified gravel

Continental glacier — moving ice sheet of continental proportions; see ice sheet

Continental rise — a broad and gently sloping ramp that rises from the deep ocean floor to the continental slope

Continental shelf — gently-sloping, submerged edge of the continent out to the continental slope, with a maximum depth of about 600 feet

Continental slope — steeper outer edge of the continent, beyond the continental shelf

Correlation — matching layers of sedimentary rock in geographically separated sections

Country rock — the host rock into which magma intrudes

Cross-bedding — sets of inclined beds formed in dunes, deltas, or beaches, especially in sand

Cross section — drawing showing features that would be exposed by a vertical slice through the Earth's crust

Crust — the outermost thin skin of the Earth made of relatively light solid rock; continental crust is mostly granitic, oceanic crust is basaltic

Crystal — geometric form of a mineral with plane faces that express an internal atomic order

Crystalline rock — igneous and metamorphic rocks

Deglaciation — uncovering of glaciated land as the ice melts

Delta — generally fan-shaped body of sediment formed at the mouth of a river where it enters standing water

Dike — tabular, igneous body formed by intrusion of magma into fractures, especially joints, that cross-cut the rock layering

Dip — angle of slope of rock layers as measured down from the horizontal

Discharge — volume of water carried through a cross section of river per unit of time, usually expressed as cubic feet per second (cfs) or cubic meters per second (cms)

Disintegration — mechanical breakdown of rock, as when water freezes in cracks and pries them apart, or rock falls from a cliff and shatters

Divide — ridge of high ground separating two streams

Dolomite — calcium magnesium carbonate mineral

Dolostone — sedimentary rock similar to limestone but composed chiefly of the mineral dolomite rather than calcite

Drift — all sediments of glacial origin

Dripstone — general term for cave deposits formed by dripping water

Drowned valley — river valley drowned when sea level rises or the land subsides

Drumlin — streamlined hill of glacial till shaped by overriding ice

Earthquake — sudden trembling of the land caused by abrupt movement along a fault, or eruption of a volcano

Epoch — geologic time unit, subdivision of a period

Era — geologic time unit including several periods

Erosion — geologic processes that wear down the land

Erratic — glacially transported boulder that generally differs from the bedrock underneath

Esker — long, low, typically sinuous ridge of sand and gravel deposited along the course of a stream that tunneled through a wasting ice sheet

Exfoliation — see sheeting

Extrusive igneous rock — volcanic rock

Fault — surface or zone of rock fracture in the Earth's crust caused by movement of one mass of rock against another

Fault scarp — cliff or steep slope on the upthrown side of a fault

Feldspar — the most common mineral of the Earth's crust; plagioclase feldspars are sodium-calcium alumino-silicates, and potassium feldspars are potassium-sodium alumino-silicates

Flour, glacial — pulverized rock produced as rocks in the ice grind against each other and against bedrock adjacent to the ice; the fine suspended material that makes meltwater streams milky and goes into the making of varved clay deposits

Flowstone — general term for carbonate deposits formed on cave walls or floor by flowing rather than dripping water

Fold — rock deformation manifested by real or apparent bending of originally planar layers

Formation — a basic unit for naming rocks, used principally for stratified rocks that are more or less homogenous and regionally mappable

Fossil — any remains, trace, or imprint of plant or animal naturally preserved in sediments or rocks from past geologic time

Frost action — mechanical weathering caused by freezing and thawing

Frost-heave — expansion and upward movement of water-soaked soil when it freezes

Gabbro — coarse-grained, dark-colored igneous intrusive rock mainly composed of plagioclase feldspar and pyroxene; intrusive equivalent of basalt

Garnet — silicate mineral common in medium- to high-grade metamorphic rocks. It forms "knots" in many schists, amphibolites, and gneisses of Vermont and New Hampshire; most garnets are red.

Glacial rebound — uplift of the land after a continental glacier melts

Glacial recession — the melting back of a glacier front primarily caused by climatic warming

Glacial scour — grinding, scraping, gouging, bulldozing action of a glacier

Glacier — mass of ice that persists through the years and flows downhill under its own weight

Gneiss — coarse-grained metamorphic rock with generally strong layering

Graben — in block-faulting, an elongate, down-dropped crustal block forming a valley, typically bounded by parallel inward-dipping faults (see Horst)

Granite — coarse-grained, igneous intrusive rock chiefly composed of potassium feldspar and quartz

Granite veining — lenses and layers of granite found in migmatites

Graywacke — "dirty" sandstone containing abundant feldspar and rock fragments, often in a clay-rich matrix

Greenschist — schist derived principally by low-grade metamorphism of basalt, and containing abundant green minerals, especially chlorite

Grenville mountain-building event — affected the east side of ancestral North America about a billion years ago as the continent collided with ancestral Africa

Grooves, glacial — large, linear furrows caused by scraping of ice-bound rocks against bedrock as the glacier moves

Groundwater — water held in pore spaces or other openings in rocks and sediments

Group — major rock-stratigraphic unit next higher in rank than formation, consisting of two or more formations

Hanging valley — small glacial valley that joins a large trunk glacial valley igh above its floor

Horst — in block-faulting, an elongate, elevated block of crust forming a ridge or plateau, typically bounded by parallel, outward-dipping faults (see Graben)

Ice sheet — thick, unconfined, sheet-like glacier, generally covering a very large area, that spreads radially over the land under its own weight; a continental glacier

Igneous rock — rock formed by solidification of magma or lava

Interglacial — period between major glaciations of the Ice Age

Intrusion — igneous rock body that forced its way in a molten state into surrounding country rock

Intrusive igneous rock — formed from molten rock that solidified below the surface

Island arc — also volcanic island archipelago; chain of volcanic islands formed at a convergent plate boundary, apparently by melting above the underthrust plate at depth

Joint — rock fracture that opens up by tension, and the opposing blocks do not slide past each other

Kame — generally conical hill of sand and gravel deposited by meltwater streams in contact with glacier ice

Kame terrace — valley side deposits of sediment laid down in a lake around the edges of a dwindling glacier tongue

Kettle — depression in glacial deposits where a buried block of ice finally melted

Klippe — isolated erosional remnant of a slice carried on a thrust fault

Law of Superposition — in undisturbed sequences of sedimentary rocks, the oldest layers are on the bottom

Lava — magma that reaches the surface

Limestone — sedimentary rock chiefly composed of the mineral calcite

Magma — molten rock that forms igneous rocks when it cools. Magma that reaches the surface is called lava.

Mantle — zone of Earth below the crust and above the core, ranging from depths of about 25 to 2088 miles

Marble — metamorphosed limestone or dolostone

Meanders — broad curves in a stream course that develop as the stream erodes its outer bank and deposits sediment against the inner bank

Meltwater — water from the melting of ice and snow, especially in glaciers

Member — subdivision of formation

Meta- (prefix) — signifies that the rock has been metamorphosed, as in metagabbro, metasedimentary, or metavolcanic

Metamorphic grade — relative intensity of metamorphism; low-grade metamorphism implies low temperatures and pressures, and so on

Metamorphic index mineral — mineral that forms or is stable over a limited range of pressures and temperatures; useful for determining conditions of metamorphism

Metamorphic rock — formed from igneous or sedimentary rocks by metamorphism

Metamorphism — changes in mineralogy and texture imposed on a rock by elevated pressure and temperature without melting

Metamorphism, contact — metamorphism caused by the heat of a nearby igneous intrusion

Metamorphism, regional — large-scale metamorphism in which elevated temperatures and pressures result mainly from deep burial, as in the core of a building mountain range; evident in nearly all of the metamorphic rocks of Vermont and New Hampshire

Mica — group of common silicate minerals that cleave easily into thin shiny sheets

Mid-ocean ridge — mid-ocean mountain range—tectonic plate boundary where plates are pulling apart

Migmatite — gneiss with intimate granitic veinings that result either from partial melting or segregation during extreme metamorphism

Mineral — natural inorganic solid with limited chemical variability and distinctive internal crystalline structure

Moraine, ground — blanket of till with no marked relief; formed under the glacier

Moraine, end — ridge of till left at the glacier front where the ice remains in fixed position for a long time in response to steady climatic conditions

Obsidian — volcanic glass formed as lava solidifies without crystallizing

Organic sediments — made up of organic material, as in coal

Outwash — glacial deposits washed over, transported, and redeposited by meltwater streams

Oxbow lake — crescent-shaped lake formed in an abandoned meander loop which has become separated by a change in the river course

Partial melt — magma produced during extreme metamorphism by preferential melting of only the rock minerals that have low melting temperatures

Pegmatite — an igneous rock with extremely large grains, more than a centimeter in diameter. It may be of any composition but most typically is granitic.

Period — geologic time unit longer than an epoch; a subdivision of an era

Phenocryst — see Porphyry

Phyllite — low-grade, fine-grained schist characterized by a pearly luster on split layers that are generally less regular than slaty cleavage

Piracy — capture of one stream by headward erosion of another leaving the lower reach of the captured stream diminished

Plastic deformation — ductile, taffy-like deformation of rocks that occurs only at depth where confining pressure and temperature are high; evident in "fluid" fold patterns

Plate tectonics — a modern field of geology which proposes that the entire crust of the Earth, including continents and ocean floors, is made up of a small number of rigid plates of enormous size that are constantly moving about

Plucking — breaking away and removal by glacial ice of large blocks from bedrock

Pluton — a large igneous intrusion, formed at depth in the crust

Polish, glacial — smooth surface on bedrock or boulder caused by the polishing action of fine-grained rock waste held by the moving ice

Porphyry — igneous rock with large crystals, called phenocrysts, "floating" in a fine-grained or glassy matrix, and common in volcanic and dike rocks

Potholes — rounded holes ground in streambed rock by sand and gravel in swirling eddies

Proglacial lake — lake formed in front of a glacier

Quartz — extremely common silicon dioxide mineral found in a wide range of igneous, metamorphic, and sedimentary rocks; the most nearly everlasting mineral of the Earth's crust

Quartzite — metamorphosed sandstone; also called metaquartzite

Radiometric clock — any of a number of radioactive elements that may be used to date the rocks that contain them

Recrystallization — a metamorphic process by which existing minerals are recrystallized or replaced by new ones without melting

Refolding — folding of rocks that have already been folded at least once, as in Acadian refolding of Taconian folds

Rejuvenation — stream erosion stimulated or renewed, as by uplift of the land or decline of sea level

Relief — maximum regional difference in elevation

Ring-dike — a dike formed where magma intruded arcuate faults caused principally by crustal collapse over the magma chamber. Many in New Hampshire are part of complexes involving two or more generations of collapse and intrusion.

River terraces — rather flat-surfaced, step-like remnants of former floodplain levels bordering a river that has cut downward

Rock — coherent granular aggregate of one or more minerals; sometimes non-mineral substances like obsidian or organic debris

Rock record — the preserved rocks

Sandstone — sedimentary rock composed of consolidated sand

Sapping — recession of a cliff by wearing away of soft layers that support a harder caprock which then overhangs and breaks off; common in the Champlain thrust fault scarp where hard quartzite overlies weak shale

Scarp — long, more or less continuous cliff or slope separating relatively flat land into two levels

Schist — generally cleavable metamorphic rock with layering defined by parallel arrangement of platy or prismatic minerals; especially mica schist

Sedimentary rock — rock formed by consolidation of sediments

Shale — sedimentary rock mainly composed of clay or clay-sized material (very fine) and often very thin-bedded

Shear zone — intensely disrupted rock zone marked by numerous closely spaced, parallel shears

Sheeting — a form of jointing most common in exposed massive igneous plutons, in which thin sheets split off more or less parallel to the surface like the layers of an onion

Shelf sequence — group of sedimentary rocks, typically comprised of carbonates, shales, siltstones and sandstones, formed from sediments deposited in the relatively shallow water environment of the continental shelf

Shield — large region of ancient basement rocks within a continent

Silica — silicon dioxide

Silicate — compound whose crystal structure contains silicate molecules. Most rock-forming minerals are silicates.

Sill — tabular igneous intrusion sandwiched between th layers of the intruded rock

Siltstone — sedimentary rock composed of silt, essentially an extremely fine sand, but coarser-grained than clay-size

Slate — low-grade metamorphic rock formed from shale and characterized by flat cleavage plates

Slaty cleavage — see cleavage, rock

Soil creep — slow downslope movement of the soil aided by alternate freezing and thawing

Strata — the layers of sediments and sedimentary rocks (also volcanic)

Stratigraphic — term used for layered sediment or sedimentary rock (also layered, or interlayered, volcanics)

Striae — linear scratches caused by movement of ice-bound rocks against each other and bedrock, often with glacial polish

Subsidence — gentle regional depression of the land surface

Superposed stream — stream that flows through resistant formations because its course was established at a higher level on uniform rocks before downcutting began; probably the best explanation for the cross-cutting courses of the Missisquoi, Lamoille, and Winooski rivers of Vermont

Syenite — coarse-grained igneous rock dominated by potassium feldspar; silica-poor granite that contains little or no quartz or plagioclase

Syncline — downfolded rock layers

Talus — fragmentary rock that accumulates at the base of a cliff, mainly by rockfall

Thrust fault — a large, low-angle push fault that transports relatively thin slices of Earth's crust long distances in response to crustal convergence

Thrust fault slice — the large, relatively thin slab of rock that moves over a thrust fault

Till — unsorted and unstratified glacial drift deposited directly from the ice without subsequent reworking by streams, usually containing material of a wide range of sizes, and often called "boulder clay"

Type exposure (or locality) — the exposure of a particular formation for which it is named, and supposedly typical of the formation

Unconformity — a surface of erosion or non-deposition separating older from younger rocks and constituting a gap in the rock record

Valley glacier — linear glacier confined to a mountain valley, also called mountain, or alpine, glacier

Varved clay — distinctly laminated clays deposited in lakes supplied by glacial meltwater streams, in which the laminations represent annual deposition like tree rings

Vein — deposit of foreign minerals within a joint or fracture

Wave-cut bench — level or nearly level surface cut by wave erosion

Wave-cut cliff — cliff produced and maintained by waves undercutting the rock or freestanding sediments along the shore

Wisconsin glaciation — the last major glaciation of the Ice Age

Xenolith — piece of country rock engulfed in intrusive or extrusive igneous rocks

Suggested Readings

Alt, David D. (1982), *Physical Geology, A Process Approach*, Wadsworth Publishing Co., 383 p.

Bennison, Adam P., compiler (1975), *Geological Highway Map of The Northeastern Region*, The American Association of Petroleum Geologists, Map No. 10

Billings, Marland P., (1955), *Geologic Map of New Hampshire* 1:250,000, jointly prepared by the New Hampshire Planning and Development Commission, The Division of Geological Sciences, Harvard University, and the U.S. Geological Survey

Billings, Marland P., (1956), *The Geology of New Hampshire Part II: Bedrock Geology*, New Hampshire Department of Resources and Economic Development, 207 p. (contains above H. H. geologic map)

Billings, Marland P. and Fowler-Billings, Katherine, (1975), *Geology of The Gorham Quadrangle, New Hampshire–Maine*, New Hampshire Department of Resources and Economic Development Bulletin No. 6, 120 p., with geologic map in pocket

Billings, Marland P., et al, (1979), *The Geology of The Mount Washington Quadrangle, New Hampshire*, New Hampshire Department of Resources and Economic Development, 44 p., with geologic map in pocket

Dodge, Harry W., Jr., (1969), *The Geology of D.A.R. State Park, Mount Philo State Forest Park, Sand Bar State Park,* Vermont Geological Survey, 32 p.

Doll, Charles G., et al, (1961), *Centennial Geologic Map of Vermont* 1:250,000, Vermont Geological Survey

Doll, Charles G., et al, (1970), *Surficial Geologic Map of Vermont,* Vermont Geological Survey

Flint, Richard F., et al, (1959), *Glacial Geology of The United States East of The Rocky Mountains (map)*, Geological Society of America

Grant, Raymond W., (1968), *Mineral Collecting in Vermont*, Vermont Geological Survey Spec. Pub. #2, 49 p.

Henderson, Donald M., et al, (1977), *Geology of The Crawford Notch Quadrangle, New Hampshire,* New Hampshire Department of Resources and Economic Development, 29 p., with geologic map in pocket

New England Intercollegiate Geologic Conference field trip guidebooks, many dealing with geology of Vermont and New Hampshire, by numerous authors, may be found in a number of New England libraries

Steward, David P. and MacClintock, Paul, (1969), *The Surficial Geology and Pleistocene History of Vermont,* Vermont Geological Survey Bulletin 31, 251 p.

VanDiver, Bradford B., (1985), *Roadside Geology of New York*, Mountain Press Publishing Co., 397 p.

Welby, Charles W., (1962), *Paleontology of The Champlain Basin in Vermont*, Vermont Geological Survey Spec. Pub. #1, 88 p.

Wilson, James R., (1969), *The Geology of The Ossipee Lake Quadrangle, New Hampshire*, New Hampshire Department of Resources and Economic Development Bulletin #3, 116 p., with geologic map in pocket

NOTE: only a few of the many publications of the Vermont Geological Survey and New Hampshire Department of Resources and Economic Development written for popular consumption are included here. Others may be obtained from the two state agencies, in Montpelier or Concord, or from some libraries.

Index